THE ART AND SCIENCE OF
WIND POWER

Wind Power Generation and Distribution

David A. Rivkin, PhD

Marc Randall

Laurel Silk, MEd, BA

JONES & BARTLETT
LEARNING

World Headquarters
Jones & Bartlett Learning
5 Wall Street
Burlington, MA 01803
978-443-5000
info@jblearning.com
www.jblearning.com

Jones & Bartlett Learning books and products are available through most bookstores and online booksellers. To contact Jones & Bartlett Learning directly, call 800-832-0034, fax 978-443-8000, or visit our website, www.jblearning.com.

Wind Power Generation and Distribution is an independent publication and has not been authorized, sponsored, or otherwise approved by the owners of the trademarks or service marks referenced in this product. Some images in this book feature models. These models do not necessarily endorse, represent, or participate in the activities represented in the images.

This publication is designed to provide accurate and authoritative information in regard to the subject matter covered. It is sold with the understanding that the publisher is not engaged in rendering legal, accounting, or other professional service. If legal advice or other expert assistance is required, the service of a competent professional person should be sought.

Production Credits

Chief Executive Officer: Ty Field
President: James Homer
SVP, Editor-in-Chief: Michael Johnson
SVP, Chief Technology Officer: Dean Fossella
SVP, Chief Marketing Officer: Alison M. Pendergast
VP, Manufacturing and Inventory Control:
   Therese Connell
VP, Production and Design: Anne Spencer
SVP, Curriculum Solutions: Christopher Will
Director of Sales, Curriculum Solutions: Randi Roger
Editorial Management: High Stakes Writing, LLC,
   Editor and Publisher: Lawrence J. Goodrich
Managing Editor, HSW: Ruth Walker
Copy Editor, HSW: Karen Annett
Senior Editorial Assistant: Rainna Erikson
Production Manager: Susan Schultz

Production Editor: Keith Henry
Production Assistant: Kristen Rogers
Senior Marketing Manager: Andrea DeFronzo
Manufacturing and Inventory Control Supervisor:
   Amy Bacus
Composition: Cenveo Publisher Services
Cover Design: Kristin E. Parker
Rights & Photo Research Associate: Lian Bruno
Rights & Photo Research Assistant: Joe Veiga
Cover Image: © Andrei Nekrassov/ShutterStock, Inc.
Chapter Opener Image: © esbobeldijk/ShutterStock, Inc.
Notes Image: © aceshot1/ShutterStock, Inc.
Tech Tip Image: © William Perugini/ShutterStock, Inc.
Warning Image: © Konstantin Yolshin/ShutterStock, Inc.
Printing and Binding: Edwards Brothers Malloy
Cover Printing: Edwards Brothers Malloy

ISBN: 978-1-4496-2450-7

Library of Congress Cataloging-in-Publication Data
Unavailable at time of publication.

6048

Printed in the United States of America
16 15 14 13 12   10 9 8 7 6 5 4 3 2 1

Brief Contents

Contents

Preface

THE WIND ENERGY INDUSTRY is at the forefront of the world's shift away from reliance on fossil fuels. In just a few short decades wind energy has evolved dramatically. Technological advances now make wind energy a cost-effective solution for the world's ever-growing energy needs. The United States is now one of the world's leaders in overall wind power capacity.

As the wind energy industry continues to expand in the United States and around the globe, it will provide many opportunities for workers in search of new careers. These careers extend beyond the wind farm and include the efforts of employees who work in manufacturing plants, offices, and construction, as well as operation and maintenance. According to estimates from the American Wind Energy Association, approximately 85,000 Americans are currently employed in the wind power industry. Despite the growing demand for skilled workers, there remains a lack of serious educational resources to meet the market's demand.

The *Art and Science of Wind Power* series was developed to fill this educational gap. Each book in the series examines performance challenges using a systems perspective. Readers do not learn design and installation steps in a vacuum—instead they examine interrelationships and discover new ways to improve their own systems and positively contribute to the industry.

This series was developed for both the novice and expert. The texts take the learner from an overview of wind energy, through design and installation steps and considerations, to the design and installation of commercial wind systems.

Wind Power Generation and Distribution provides readers with information on electric motors and the installation and maintenance of wind turbines. Topics include energy conversion, power electronics, converters, generators, wind-turbine control, rotor dynamics, and wind farms.

About the Authors

Prof. David A. Rivkin, PhD, is managing director and dean of the College of Science and Technology at the Sustainable Methods Institute (SMI), an online university and innovation center.

He is also the chairman of the Department of Nanosciences in Renewable Energy at Chiist University, Atlanta, Georgia; and dean of education and research at the Israel Sustainability Institute. He is also chief scientist and director at the Adamah Group and its wind power division in Israel.

Professor Rivkin was the founder, associate professor, and chairman of the Green Technologies Department at Ohalo College of Katzrin; associate professor at National University, based in San Diego, Calif.; International Technological University, in San Jose, Calif.; and the Graduate School of Science and Technology at the Technical University of Munich, in Germany.

He holds dual bachelor's degrees in chemistry and nuclear engineering from the University of California at Berkeley. Professor Rivkin pursued postgraduate studies toward a doctorate in biophysics at the University of California and later completed a PhD in business sciences at the European School of Business London at Regent's College, with a focus on small business sustainability. He has over 25 years of professional experience in both industry and academia.

He has taught at internationally renowned colleges and universities in Europe, India, China, the United States, and Israel. In 2010 Professor Rivkin was nominated to be a Fulbright scholar. He is also an Institute of Electrical and Electronics Engineers (IEEE) senior member and distinguished lecturer, a principal adviser in clean technologies to the National Science Foundation, a certified program manager, a certified corporate sustainability expert, and a certified green energy professional.

He is the winner of numerous technical and managerial awards and has been recognized for his outstanding contributions by governments as well as the United Nations. A serial entrepreneur, with roles in several successful ventures, including as founder and chief of technology for SciEssence International, Professor Rivkin has a multidisciplinary background, from biosciences to nanotechnology from health to energy, that gives him broad expertise in sustainability.

Marc Randall is a freelance writer and instructional designer currently living in Buenos Aires, Argentina. Born in New Mexico, he studied education at Colorado State University in Fort Collins, Colorado. He has been involved in print-based and online education for nearly a decade.

Laurel Silk has managed e-learning initiatives at three leading universities in Arizona including Arizona State University, University of Phoenix, and Grand Canyon University. Among the highlights of her career, she created a virtual doctoral library for research students and designed and implemented a web-based doctoral studies program in administration. Ms. Silk is a former classroom instructor, in which capacity she designed and taught courses in freshman English, world literature, and critical thinking. As vice president of SilkWeb, she has created undergraduate and graduate online courses in higher education, renewable energy, business, and nursing. Ms. Silk holds a Masters degree in education with a focus on adult learning and instructional design technologies from University of Phoenix and has a Bachelor of Arts in English from Arizona State University.

Introduction: A History of Wind Power

FOR OVER 5,000 YEARS, wind has been a viable source of power, generating energy for a variety of purposes, ranging from sailboats to machinery. Wind power has been converting wind energy into useful and renewable energy as wind turbines for electricity, windmills for mechanical power, and wind pumps for water pumping or draining. The use of **wind** for sailing later inspired Greek engineer Heron of Alexandria to develop the first wind-powered machine in the first century AD—an organ **FIGURE 1**. Other ancient uses of wind include the prayer wheels of Tibet and China in the fourth century. Centuries followed before the first applied technology in wind power emerged with the invention of the windmill in the seventh to ninth centuries. The first practical windmill was built in Sistan (now eastern Iran); it was composed of a long, vertical **drive shaft** with rectangular blades and reed- or cloth-covered sails **FIGURE 2**. Though able to grind corn and pump water, this vertical design proved to be inefficient and highly suscep-tible to damage as the vertical axes of the rotors were driven by drag forces.

NATURE USES WIND TO POWER TRANSPORT

Nature has been using the power of the wind for millions of years to transport pollen and seeds to distant locations. Dandelion plants evolved specific lightweight seeds with parachute-like tops to catch the wind and be blown miles from their parent. Maple trees evolved pro-peller-like wings on their seeds not only to drop safely to the ground but

(Continues)

NATURE USES WIND TO POWER TRANSPORT (Continued)

also to be caught by the wind and blown far from the parent tree. Pollen grains are microscopically shaped to catch the wind and be carried for many hundreds of miles. From nature, we can learn a great deal about using the wind to provide us with power in many forms.

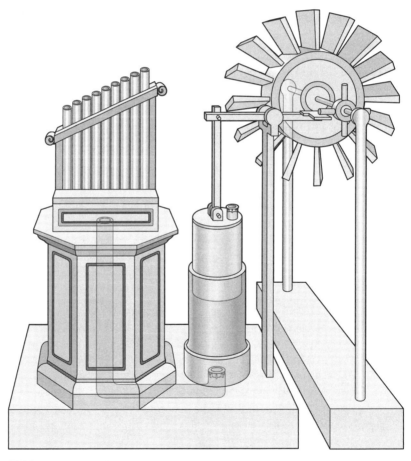

FIGURE 1   Heron's wind-powered organ.

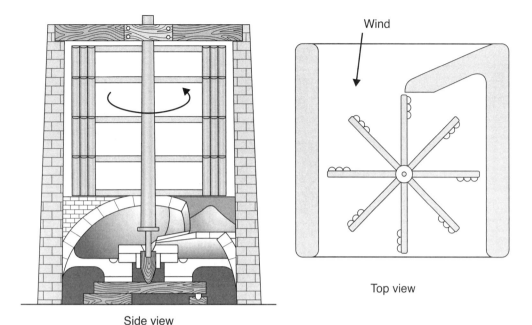

Wind

Top view

Side view

FIGURE 2 The top and side view of the Sistan windmill.
Courtesy of Kaboldy

It was not until the early twelfth through fourteenth centuries that Europe began using horizontal-axis windmills, which greatly improved the performance and maintenance of wind energy generation. Unlike water mills, wind-powered mills were not restricted to seasonal changes, such as water freezing in winter, and were not confined to the proximity of fast water streams. With about four blades, the first version of the European windmill was the post mill, which turned to face the wind (or **yaw**) depending on wind direction. Later versions (smock mills), as shown in **FIGURE 3**, improved the European design by keeping most of the mill stationary and making only the top portion movable to face the wind.

Further advancements in windmills in the eighteenth century spanned research and design into theory and development. In England, John Smeaton developed the three basic principles of windmill design that are still relevant today:

- The speed of the blade tips is ideally proportional to the speed of wind.
- The maximum torque is proportional to the speed of wind squared.
- The maximum power is proportional to the speed of wind cubed.

As uses for the windmill increased (e.g., pumping water for salt making), the demand for windmills reached large numbers in the late 1800s, with Denmark

FIGURE 3 Smock mill in Amsterdam.
Courtesy of Aloxe Alix Guillard

establishing 2,500 windmills to offset their below-sea-level terrain, which was subject to frequent flooding. During the latter half of the nineteenth century, the American Midwest alone became the home to approximately 6 million "fan mills," which provided advanced regulated irrigation (unattended for long periods) to small farms.

Beyond irrigation purposes, wind energy became part of a new branch of study—electrical science and engineering—at the advent of the Second Industrial Revolution. In 1887, Professor James Blyth of Scotland and Camille Fauré of France supplied the first wind-powered electricity to a cottage in Marykirk, Scotland. That same year in Cleveland, Ohio, Charles Brush built a massive wind-powered machine with a turbine **rotor** spanning 56 feet in diameter that was mounted on a 60-foot tower **FIGURE 4** . With 144 blades, this slow machine had an output of a mere 12 kilowatts (kW), which could power only about 100 light bulbs, three arc lamps, and some motors.

Electricity-generating wind turbines gained more momentum in the early 1900s. Based on the smock mills, Danish scientist Poul la Cour constructed over 100 electricity-generating turbines by converting the produced electricity to hydrogen gas, which was then used for lighting. Then, in 1927, the Jacobs brothers, Joe and Marcellus, became the first successful commercial producers of **wind**

FIGURE 4 The Brush windmill.

turbine generators for agricultural purposes, with about 30,000 small wind turbines in the United States. A few years later in 1931, G. J. M. Darrieus invented a
new wind turbine **FIGURE 5** with a vertical-axis eggbeater design, allowing wind
from any direction with no need for adjustment. By the 1930s, the two-blade, horizontal-axis Wincharger turbine became the most widely used small wind

FIGURE 5 The Darrieus windmill.

FIGURE 6 The Smith-Putnam windmill.
Courtesy of NASA

generator, outputting up to 200 watts. A decade later, the world's first megawatt wind turbine was invented by Palmer Putnam and S. Morgan Smith. The Smith-Putnam turbine **FIGURE 6** produced 1.25 megawatts (MW) and operated for 1,100 hours.

LIMITED RESEARCH INTO VERTICAL DESIGNS

Since the 1970s, wind turbines with a vertical axis of rotation have received little technical attention and research, until recently. The reason for this is not that vertical designs are inherently limited. Rather, this lack of attention goes back to the simple mechanical failure of a support wire during an important demonstration of one Darrieus windmill for the US Department of Energy. The false perception that vertical designs are mechanically not as viable has hindered their development.

FIGURE 7 The experimental Mod-5B wind turbine on the Hawaiian island of Oahu.
Courtesy of NASA

Forty years passed with the Smith-Putnam wind turbine technology in the lead and no new wind energy development. Then, in the mid-1970s, the oil price crisis engendered a reemergence of wind energy investigation. Environmental concerns revolving around fossil fuels and potential dangers of nuclear energy drove more research into "alternative" forms of energy, including wind power. The US Department of Energy (DoE), NASA, and the National Science Foundation pooled efforts into multimegawatt turbine technologies (Mod-series). The private turbine manufacturing industry adopted many of their findings, so much of their work is still valuable today. In 1981, the Mod-2 wind turbine cluster produced 7.5 MW and in 1987, Mod-5B **FIGURE 7** became the world's largest single wind turbine to produce 3.2 MW independently. With a 100 meter (m) diameter, sectioned two-blade rotor, and 95 percent availability, the Mod-5B had the world's first large-scale variable-speed drive train. California pushed the wind energy movement further by providing tax rebates for wind power in the 1980s, resulting in a boom of wind farms **FIGURE 8**.

Today, fossil fuel dependency, energy security, and global warming are still relevant and prevalent concerns, stimulating great interest in wind power as a genuine source of alternative energy. In 2006, a $10,000 wind power unit could

FIGURE 8 Wind farms in California.
Courtesy of Stan Shebs

generate 80 percent of the average home's power needs. In 2008, Rock Port, Mo., became the first US city to run on 100 percent wind power, courtesy of the Wind Capital Group. Then, in 2009, the United States overtook Germany as the world's leader of installed wind power, with about 2 percent of national power originating from wind turbines. Wind power capacity reached over 9,900 MW in the United States in 2009. That much capacity can easily power 2.4 million homes. At this rate, wind energy could account for 20 percent of American electricity by the year 2030. As one solution to environmental hazard concerns, fossil fuel depletion, and the need for energy security, the advancement of wind energy generation presents an opportunity. For a brief historical timeline of wind energy, see **TABLE 1**.

TABLE 1 A BRIEF TIMELINE OF WIND ENERGY	
Year	**Event**
100s	Heron of Alexandria develops the first wind-powered machine, an organ.
300s	Prayer wheels of Tibet and China are driven by wind.
600s	The first practical windmill is built in Sistan (eastern Iran).

Year	Event
TABLE 1 A BRIEF TIMELINE OF WIND ENERGY (Continued)	
1100s	The first Dutch windmills are developed and begin converting wind energy to mechanical energy.
1300s	Smock mills improve on previous European designs, featuring movable tops that allow optimal wind energy extraction.
1700s	Windmill design principles are developed by John Smeaton.
1800s	Approximately 6 million American fan mills begin providing advanced regulated irrigation to small farms.
1887	Blyth and Fauré supply a house with the first wind-powered electricity in Scotland.
1887	Brush builds the first wind-powered turbine to output 12 kW in Cleveland, Ohio.
1900s	Poul la Cour constructs electricity-generating turbines for lighting.
1927	The Jacobs brothers become the first successful commercial wind turbine producers.
1930s	The Wincharger turbine with a two-blade horizontal axis becomes popular in America's heartland.
1931	Darrieus invents a new wind turbine with a vertical-axis eggbeater design.
1940s	The Smith-Putnam turbine becomes the world's first megawatt wind turbine.
1970s	The oil price crisis spurs the reemergence of wind energy development.
1980s	California provides tax rebates for wind power and the wind farm boom begins.
1981	The Mod-2 wind turbine cluster produces 7.5 MW.
1987	The Mod-5 becomes the world's largest single wind turbine to produce 3.2 MW alone.
2006	A $10,000 wind power unit can generate 80% of the average home's power needs.
2008	Rock Port, Missouri, becomes the first US city to run on 100% wind power.
2009	United States surpasses Germany as the world's leader in installed wind power capacity.
2030	Wind energy may account for 20% of American electricity.

KEY CONCEPTS AND TERMS

Drive shaft	Wind
Generator	Wind turbine
Rotor	Yaw

Wind Energy Generation and Conversion

PEOPLE HAVE BEEN HARNESSING the power of the wind for thousands of years—to sail across the seas, to turn windmills, and to pump or drain water, among other uses. Nowadays, wind power is of particular interest as a way to generate electricity. In this chapter, you will read about more recent developments in wind power, and begin to get a grasp of the basics of wind turbines and their functions.

Chapter Topics

This chapter covers the following topics and concepts:

- How wind turbines work
- Type and size of wind turbines
- Inside a working wind turbine
- Wind energy potential in the United States
- Wind farms overview
- Wind energy generation

Chapter Goals

When you complete this chapter, you will be able to:

- Understand how wind is converted into energy
- Discuss the energy potential available in the United States
- Briefly describe how a wind farm works when connected to the grid
- Briefly describe the components of a wind turbine engine

How Wind Turbines Work

A wind turbine is a wind-powered electrical machine that uses wind to drive an electric generator. The turbine uses **kinetic energy** derived from wind. Wind passes over the rotor blades to generate lift and exert a turning force. The produced airfoils transform the kinetic wind energy into mechanical energy. This energy is then converted into an electrical energy, which can be interconnected to a power grid or used on-site.

The Definition of Wind

To understand how a wind turbine works, you should understand the dynamics of wind. Wind is air in motion, in any direction. As a type of solar energy, wind is caused by spatial differences in atmospheric pressure, specifically moving from high pressure to low pressure. Earth is unevenly heated by the Sun; that is, the North Pole and the South Pole receive less solar energy than the equator, and dry land heats and cools more quickly than the seas. This difference in heating causes global atmospheric convection—air flowing from high pressure to low pressure—which in turn causes wind. The direction of wind is influenced by Earth's rotation. Wind turbines use wind or the motion of air (*flow field*) to power machines and/or produce electricity.

Wind Turbine Revealed

A wind turbine produces energy by harnessing the kinetic energy of wind and converting it into useful energy, for example, electrical energy. Wind moves the blades of a wind turbine, which in turn spins the shaft that is connected to a generator to produce electricity. As the wind passes over the blades, it creates a lift. As the blades rotate, a shaft inside the nacelle (the section of the turbine that contains the gearbox, shafts, generator, controller, and brake) turns. Some nacelles also contain gearboxes. They increase the rotational speed so that the generator can convert the rotational energy into electrical energy. The power generated is delivered to an interconnection substation via a collection system. At the substation, power is typically transformed to a higher voltage before delivery to the power grid.

Calculating Wind Turbine Power

Basic wind turbine technology includes the conversion of wind (kinetic energy) into mechanical power or electricity. Wind is made of moving air molecules, which have mass; any moving object with mass has kinetic energy. Many factors must be considered in calculating wind's electrical generation capability. However, you only need to understand the following formulas. The power theoretically available is related to the following:

- The area swept by the rotor, so if you double the swept area, the power output will also double.

- The cube of the wind speed, so if the wind speed is doubled, the power output will increase by a factor of 8 (or 2^3). In general, the higher the tower is off the ground, the more power it can produce, because wind speeds increase roughly exponentially with height.

The formula for the kinetic energy of a moving fluid is as follows:

$$\text{Kinetic energy} = 0.5 \times \text{Mass} \times \text{Velocity}^2$$

where kinetic energy is measured in joules and mass is measured in kilograms. Velocity is measured in meters per second.

The mass of air contacting a wind turbine (which sweeps a known area) per unit of time is calculated as follows:

$$\text{Mass/sec (kg/s)} = \text{Velocity (m/s)} \times \text{Area (m}^2) \times \text{Density (kg/m}^3)$$

Then, if the preceding formula is inserted into the aforementioned kinetic energy formula, the following vital equation results:

$$\text{Power} = 0.5 \times \text{Swept area} \times \text{Air density} \times \text{Velocity}^3$$

Here, you express power in watts (joules/second), the swept area in square meters (m^2), the air density in kilograms per cubic meter (kg/m^3), and the velocity in meters per second (m/s).

For example, the world's largest wind turbine (its rotor blade diameter is 126 m) sweeps an area of 12,470 m^2. Because this turbine is offshore at sea level, the air density is 1.23 kg/m^3; thus, if rated at 5 MW in 30 miles per hour (mph) winds (14 m/s), the following can be determined:

$$\text{Wind power} = 0.5 \times 12{,}470 \times 1.23 \times (14^3) = 21{,}000{,}000 \text{ watts!}$$

One million watts is equal to 1 megawatt (MW), so you can express the final result as 21 MW. How can the power of the wind (21 MW) be so much larger than the rating of the turbine generator (5 MW)? The Betz limit explains this phenomenon.

Betz Limit

In 1919, German physicist Albert Betz stated that no wind turbine could convert more than 16/27 (59.3 percent) of the kinetic energy of wind into useful energy; this law is known as the **Betz limit**. Wind turbines pull energy essentially by slowing the wind. To extract 100 percent of the wind energy, it would need to stop 100 percent of the wind, but then the rotor would not turn. See **FIGURE 1-1** for a diagram illustrating the Betz limit.

Theoretically, the instantaneous maximum power efficiency of any wind turbine design is 59.3 percent, due to the Betz limit. Due to this and other practical limitations, only 10 to 30 percent of the actual wind energy is collected and converted into useful energy.

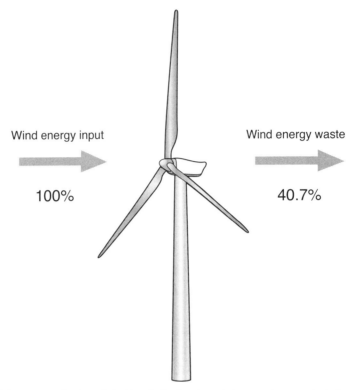

Wind energy input

100%

Wind energy waste

40.7%

FIGURE 1-1 Visualization of the Betz limit, showing wind energy input and wind energy waste.
Adapted from http://www.reuk.co.uk/Betz-Limit.htm

Type and Size of Wind Turbines

A wind turbine can vary in type and size, depending on use. A wind turbine could be as small as a 100 kW home generator or a 650-foot Enercon E-126 with a rated capacity of 7.58 MW. As for type, they can be either a **horizontal axis wind turbine (HAWT)** or a **vertical axis wind turbine (VAWT)**. HAWTs are more common than VAWTs, as theoretically they have higher power efficiencies. However, VAWTs do not depend on wind direction, so time and power are not wasted in searching for wind. Wind direction does change constantly in some environments. Where this happens, in theory, VAWTs can generate electricity more efficiently than some HAWTs. The one commonality to all turbines is the shutdown feature to avoid catastrophic damage at high wind speeds. There are many VAWT models, but in comparison to HAWTs, they are very rarely used in commercial generation. HAWTs are much more commonly used in wind energy production.

Horizontal Axis Wind Turbine (HAWT)

Horizontal axis wind turbines (HAWTs) usually have three blades that operate upwind (toward wind). The main rotor shaft and electrical generator are located at the top of the tower and face upwind. Many small HAWTs use a wind vane (weather vane) and large turbines use wind sensors (with a servo motor). Most horizontal axis turbines have a gearbox to increase the rotational speed of the generator well above the rotational speed of the blades. The turbine must point upwind because the structure itself produces turbulence behind it. To prevent damage, the turbine blades of a HAWT are usually very sturdy and positioned as far away from the tower as possible (sometimes tilted forward at the lower arc). Though usually able to produce more energy than their counterparts (VAWTs), horizontal axis wind turbines employ automated systems that readjust the orientation of the nacelle such that the rotor remains perpendicular to oncoming wind. Downwind HAWTs have been manufactured to reduce wind realignment and decrease blade damage. These HAWTs have flexible blades, because there is no risk of tower interference, but the flexible blades can reduce the swept area. Wind turbulence causes fatigue failure in most downwind turbines. Thus, to reap the most benefits, most horizontal axis wind turbines tend to be upwind rather than downwind.

Most modern wind turbines on wind farms are horizontal axis with three blades and face upwind. These highly efficient and highly reliable (low torque ripple) commercial turbines reach high tip speeds of over 320 kilometers per hour (kph). Blades that range from 20- to 40-plus meters rotate at 10–22 revolutions per minute. Knowing this information is helpful for finding **tip–speed ratio (TSR)**. TSR is the ratio between the blade tip speed and the current wind speed in a given moment. If the tip speed is exactly the same as the wind speed, TSR is 1. TSR is related to efficiency. Higher tip speeds result in higher noise levels and, due to large centripetal forces, the need for stronger blades. The tubular steel towers of these turbines vary in height from 60 to 90 meters. Gearboxes are usually used to adjust the speed of the generator; otherwise, annular generators (direct drive) are used, which negates the necessity of a gearbox. Most of these turbines are variable-speed types that use solid-state electronic power converters to more efficiently collect energy from the wind. The advantages and disadvantages of horizontal axis wind turbines are further detailed in **TABLE 1–1**.

Vertical Axis Wind Turbine (VAWT)

Vertical axis wind turbines (VAWTs) are advantageous in high-turbulence sites where wind direction is highly variable, due to the vertical main rotor shaft design. Because the generator and gearbox are placed near the ground, VAWTs do not require a tall tower, and maintenance is relatively easy. However, the intrinsic vertical design creates pulsating torque and less wind energy collection. Because the

TABLE 1-1 ADVANTAGES AND DISADVANTAGES OF HORIZONTAL AXIS WIND TURBINES

Advantages	Disadvantages
Remote variable blade pitch allows optimal turbine blade angle of attack and greater control to capture maximum wind energy at any time, day, or season.	Large towers and blades involve high costs for transportation and installation.
The tall tower base permits greater energy capture at greater height due to wind shear (increase of wind speed with height).	Reflections from tall HAWTs can create radar signal clutter.
Perpendicular movement of blades relative to wind velocity allows the whole rotation to absorb power so it is highly efficient. Airfoils never have to travel against wind.	Local opposition (landscape disruption) is caused by high visibility of tall HAWTs.
HAWTs provide consistent lateral wind loading, reduced rotor vibration, and less noise pollution because their horizontal-axis blades face wind at a consistent angle.	Downwind HAWTs are more prone to fatigue and structural failure.
	Additional yaw control is required to turn the blades and nacelle toward the wind.
	There is a higher potential for cyclic stresses and vibrations that fatigue the blade and axle.

turbine is closer to ground level, wind speed is slower at the lower altitude and turbulence is greater (causing greater vibration).

Vertical axis wind turbines are categorized in three subtypes: Darrieus, Giromill, and Savonius. Frenchman Georges Darrieus invented the Darrieus design ("eggbeater" shape), which possesses good efficiency but poor reliability (high torque ripple and cyclical stress). Also, because the starting torque is so low, the Darrieus wind turbine typically requires external power. These wind turbines also require enhanced rigidity by use of an external superstructure connected to the top bearing.

Giromills are a cycloturbine that is similar to a Darrieus turbine except the blades are straight, which allows for variable pitch to reduce torque pulsation and allow self-initiation. Savonius wind turbines are drag-types with two or more scoops used in anemometers. Anemometers are small turbines used to measure wind speed. Savonius turbines are also used in Flettner vents (used on bus and van roofs). If three or more scoops are used, the turbine is self-starting, and if the scoops are long and helical, a smooth torque is achieved.

Inside a Working Wind Turbine

Most wind turbines are composed of similar components. This chapter discusses the inside of a conventional horizontal axis wind turbine. To exploit wind energy at any given location, three basic components are required: the rotor, generator, and structural support.

Rotor

The rotor includes the blades and the hub together. Normally, it accounts for about 20 percent of the turbine's cost. Most turbines have either two or three blades that allow lift and rotation. The rotor turns the low-speed shaft at ~30 to 60 rotations per minute (rpm). The rotor speed is controlled by the blades that are turned (or pitched) in relation to the velocity of the wind.

Generator

In the past, generators were usually of a standard induction design and produced alternating current (AC) electric power. Nowadays, many generators use a variable-speed and variable-frequency design. The high-speed shaft drives the generator. The generator is contained in the **nacelle** (atop the tower). The nacelle also houses the gearbox, low- and high-speed shafts, controller, and brake. The **gearbox** connects the low-speed shaft to the high-speed shaft and increases the rotational speeds from 30–60 rpm to 1,000–1,800 rpm to produce electricity. The **controller** starts the turbine at 8–16 mph and shuts down at 55 mph to prevent high wind damage. The **disk brake** (mechanical, electrical, or hydraulic) stops the rotor in emergency situations. Outside of the nacelle is the anemometer and wind vane. The **anemometer** measures wind speed and transmits wind speed data to the controller. The **wind vane** measures wind direction and communicates with the yaw drive to manipulate the turbine according to wind orientation.

Structural Support

The structural support contains the tower, yaw drive, and yaw motor. The tower is usually made from tubular steel or steel lattice. The taller the tower, the more energy a given turbine may produce due to wind shear. The **yaw drive** keeps the rotor facing into the wind whenever wind direction changes; upwind turbines require a yaw drive to face into the wind. Downwind turbines require no yaw drive. The **yaw motor** powers the yaw drive.

Wind Energy Potential in the United States

In theory, wind energy production in the United States could satisfy 100 percent of national electricity demand. However, conditions like changeover costs and the lack of enough suitable wind environments make this impractical. With rising concerns of fossil fuel depletion, nuclear production controversy, and traditional energy production's environmental effects, wind power will continue to grow. How much production can the United States realize?

In 2008, the US Department of Energy (DoE) published the results of a study conducted to find out how feasible it would be to generate 20 percent of the United States demand for electricity by 2030. The study found that if wind energy is to meet 20 percent of the United States electricity demand by 2030:

- There is a need for enhanced transmission infrastructure as well as streamlined site placement and government regulation.
- Wind systems' reliability and operation efficiency must increase.
- The number of turbine installations must increase from about 2,000/year (2006) to almost 7,000/year by 2017.
- Twenty percent capacity integration costs should end up being less than $0.05/kilowatt-hour (kWh).
- This goal would not be limited by raw material availability.
- Additional transmission and distribution resources must be installed in parallel with excess wind energy generation capacity.

As you can see, achieving 20 percent market share by 2030 is not a foregone conclusion, but it is a possibility. What you can count on is that even if penetration does not reach 20 percent by 2030, wind energy production will keep growing in the United States and globally for years to come.

Wind Farms Overview

A wind farm is a collection of wind turbines in the same location used for the production of electricity. Single turbines are interconnected to a medium-voltage (~34.5 kilovolt [kV]) power collection system. Using a transformer, this medium-voltage electrical current is then transformed into high voltage at a nearby substation in order to connect with a high-voltage transmission. In some areas, wind farming systems can cover hundreds of square kilometers and the land between groupings of turbines is used for agriculture or other purposes. Usually, wind farms contain a few dozen to several hundred wind turbines and can be found either on land or offshore.

Several factors are analyzed in determining good wind farm sites. One factor is **wind power density (WPD)**, which is the calculation of effective wind force (velocity and mass) in a particular location, at a given height above ground level, over a specified time period. When the WPD increases, the class rating increases. Wind power density is affected by wind speed, altitude, and the "wind park effect" (loss of output due to mutual interference among turbines). Other factors include environmental aesthetics, bird life endangerment, noise pollution **FIGURE 1-2** , and power grid accessibility. The farther away the wind farm is from the power grid, the more transmission lines are required to transfer the harnessed wind power.

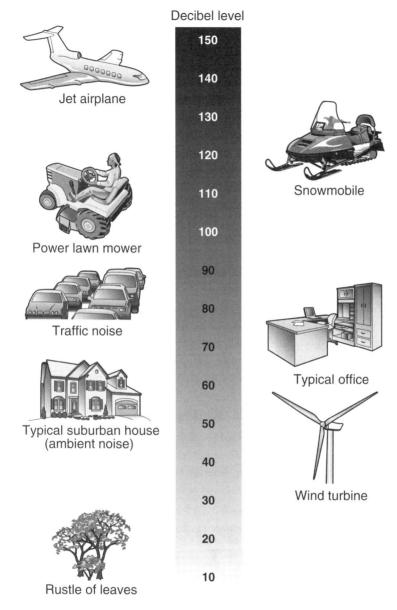

Decibel level

Jet airplane

Power lawn mower

Traffic noise

Typical suburban house
(ambient noise)

Rustle of leaves

Snowmobile

Typical office

Wind turbine

FIGURE 1–2 Noise level of an average wind turbine.
Adapted from http://www.totalalternativepower.com/faq.htm

Wind farms are also developed **onshore**, **nearshore**, or **offshore**. Onshore turbines tend to be in higher terrain than surrounding areas because wind accelerates over ridges, and often wind turbines can produce more power if they are higher up. If onshore turbines are near a coast, they are usually at least 3 kilometers (km) from it.

Nearshore turbines are usually located closer to the shore (within 3 km) or on water within 10 km of land; convection increases wind speeds as the heating of land and sea differentiate. Offshore farms are at least 10 km away from land and the distance reduces noise pollution and aesthetic environment issues; however, utilization rates, transportation, installation, and maintenance costs are much higher. Offshore wind farm technology involves either a fixed-bottom, foundation-based tower or a deep-water, floating turbine.

Between onshore and nearshore wind farms, the United States leads the world in installed wind energy capacity, whereas Spain, Denmark, and Germany lead European wind energy development. According to the American Wind Energy Association, the United States produced 21,000 MW of wind energy capacity in 2009. This energy capacity can serve 5 million average households. The Roscoe Wind Farm in Texas was claimed to be, at its completion in 2009, the largest-capacity wind farm in the United States and the world, outputting 780 MW. California leads the nation in the number of wind farms and variety of turbine types as well as housing the earliest and largest American wind farms, despite the lack of commercially viable wind farm sites onshore. The Great Plains and Texas are more conducive to wind energy developments. The United States' first offshore wind farm was approved by government regulators in 2010. It is the Cape Wind project in Massachusetts' Nantucket Sound. There are about a dozen other proposed offshore wind projects in the United States. All of the rest are still waiting for final government approval.

Wind Energy Generation

In 2009, wind energy accounted for less than 3.6 percent of the US electricity generation FIGURE 1-3. Globally, renewable energy generation is at 18 percent of total. Of that 18 percent, hydroelectricity generates 15 percent and other renewables, including wind, generate 3 percent. However, wind power generation is growing at 30 percent each year, and in 2009 there was a total wind generation capacity of

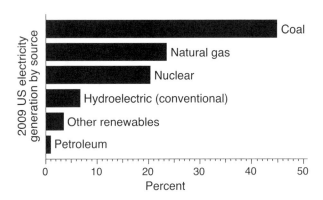

FIGURE 1-3 American electricity generation by source in 2009

Data from http://en.wikipedia.org/wiki/File:2008_US_electricity_generation_by_source_v2.png

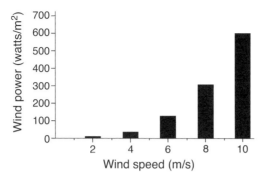

FIGURE 1-4 Wind speed versus wind power.

Data from http://www.totalalternativepower.com/faq.htm

158 gigawatts (GW). Wind energy is used mostly in Europe, Asia, and the United States. Producing no greenhouse gases (e.g., carbon dioxide and methane) during operation, wind power is the most noninvasive and viable form of renewable alternative energy source.

In generating wind energy, the relationship between wind speed and wind power is cubic **FIGURE 1-4**. Wind speeds increase roughly exponentially with height. Thus, tower height can dramatically increase the amount of electricity produced. Once wind energy is harnessed, the energy is transferred to a power grid, as shown in **FIGURE 1-5**.

NOTE

It is predicted that in 2012 China will exceed the production of electricity from wind power of all other countries combined.

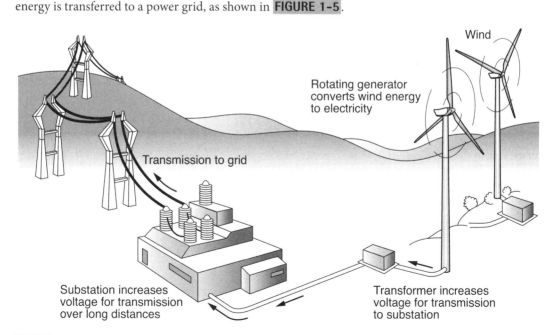

FIGURE 1-5 Diagram of how energy is transferred from a wind turbine to the power grid.

Adapted from GreenPowerOregon.com

BENEFITS OF 20 PERCENT WIND ENERGY BY 2030

According to the US Department of Energy, the United States has a wind energy potential to reach 20 percent of the nation's energy supply by the year 2030. The emphasis for greater energy efficiency with alternative and renewable energy projects gains increasing momentum in today's energy fossil fuel and environmental concerns. Increasing wind energy consumption to 20 percent by 2030 has other indirect economic impacts, such as producing half a million wind industry jobs and increasing tax revenues by $1.5 billion. A diverse energy portfolio guarantees energy security and electric price stability as well as reduction in greenhouse gas emissions by about 25 percent. Despite higher initial capital costs, wind energy could potentially offset $155 billion by lowering fuel expenditures. For more wind energy fact sheets, see the American Wind Energy Association website at *www.awea.org/pubs/factsheets.html* or the 20 Percent Wind Project at *www.20percentwind.org/.*

CHAPTER SUMMARY

For thousands of years, mankind has utilized wind for technological advancements in sailing, water pumping, grain grinding, operating machinery, and generating electricity. Wind turbines have evolved from the antiquated Dutch smock mills to the latest designs, capable of outputting several hundreds of megawatts. They can do this because of technological advances found along each step of the evolutionary path and applied in greater numbers on each subsequent turbine design. Nowadays, two- to three-blade horizontal-axis wind turbines fill wind farms whose growth has boomed in the United States and overseas.

Since the 1980s, wind has been one of the world's fastest growing renewable energy technologies. Currently, the United States leads the world in wind energy generation, with about 1 percent of the total energy capacity utilized. The US Department of Energy projects 20 percent of electricity consumption could be drawn from wind energy by the year 2030. Diversity in energy sources helps contribute to the United States' independence from fossil fuels. It also helps to protect the environment from excess pollution.

CHAPTER KEY CONCEPTS AND TERMS

Anemometer

Betz limit

Controller

Disk brake

Gearbox

Horizontal axis wind turbine (HAWT)

Kinetic energy

Nacelle

Nearshore

Offshore

Onshore

Tip–speed ratio (TSR)

Vertical axis wind turbine (VAWT)

Wind power density (WPD)

Wind vane

Yaw drive

Yaw motor

CHAPTER ASSESSMENT: WIND ENERGY GENERATION AND CONVERSION

1. A disk brake can be applied in the following ways:
   - ❑ **A.** mechanically.
   - ❑ **B.** electrically.
   - ❑ **C.** hydraulically.
   - ❑ **D.** All of the above

2. What is wind energy?
   - ❏ **A.** The process by which wind is used to generate mechanical power or electricity.
   - ❏ **B.** The process by which wind is used to engineer mechanical power or electricity.
   - ❏ **C.** The process by which wind is used to convert mechanical power or electricity.
   - ❏ **D.** The process by which wind is used to mechanize renewable electricity.

3. By 2030, the US Department of Energy would like to have achieved which goal?
   - ❏ **A.** 20 percent of the world energy consumption to be wind energy
   - ❏ **B.** 15 percent of the national energy consumption to be wind energy
   - ❏ **C.** 20 percent of the national energy consumption to be wind energy
   - ❏ **D.** None of the above

4. What are two advantages of wind-generated electricity? (Select two.)
   - ❏ **A.** Free, renewable resource
   - ❏ **B.** Clean, nonpolluting electricity source
   - ❏ **C.** Greenhouse gas resulting from pollutants
   - ❏ **D.** Onshore development of fossil fuels

5. Noise pollution is a disadvantage of wind turbines.
   - ❏ **A.** True
   - ❏ **B.** False

6. Darrieus wind turbines are a type of _____.

7. Horizontal axis wind turbines typically have two or three blades.
   - ❏ **A.** True
   - ❏ **B.** False

8. Which of the following definitions best defines tip-speed ratio (TSR)?
   - ❏ **A.** TSR is the ratio between the blade tip speed and the current wind speed averaged over a period of time. If the tip speed is exactly the same as the wind speed, TSR is 1.
   - ❏ **B.** TSR is the ratio between the blade tip speed and the current wind speed in a given moment. If the tip speed is roughly the same as the wind speed, TSR is 1.
   - ❏ **C.** TSR is the ratio between the blade tip speed and the current wind speed in a given moment. If the tip speed is exactly the same as the wind speed, TSR is 10.
   - ❏ **D.** TSR is the ratio between the blade tip speed and the current wind speed in a given moment. If the tip speed is exactly the same as the wind speed, TSR is 1.

9. Which of the following is *not* a component of the nacelle, which sits on top of the tower?
   - ❏ **A.** Gearbox
   - ❏ **B.** Yaw drive
   - ❏ **C.** Generator
   - ❏ **D.** Blade

10. Wind turbines are *not* hazardous to birds and bats.
    - ❏ **A.** False
    - ❏ **B.** True

11. Which of the following items do *not* necessarily have to be installed on a wind farm?
    - ❏ **A.** Generators
    - ❏ **B.** Substations
    - ❏ **C.** Towers
    - ❏ **D.** VAWTs

12. In generating wind energy, the relationship between wind speed and wind power generated is _____ and _____. (Select two.)
    - ❏ **A.** directly proportional
    - ❏ **B.** a function of swept area
    - ❏ **C.** dependent on blade construction material
    - ❏ **D.** cubic.

13. An anemometer performs the following function:
    - ❏ **A.** stops a motor in an emergency.
    - ❏ **B.** connects a low-speed shaft to a high-speed shaft.
    - ❏ **C.** measures wind speed.
    - ❏ **D.** drives the generator.

Modern Power Electronics and Converter Systems

IN THIS CHAPTER, you will read about the general principles behind modern power electronics and converter systems. The chapter opens with a short discussion of wind forecasting. You will read about the methods behind wind forecasts and explore what wind forecast models are based on.

The chapter then moves on to discuss electronic devices. You will cover electromechanical relays as a basis for the formation of other more complex electronic devices. These include contactors, controllers, and switches. After you conclude electronic devices, you will then cover power converters. These are devices that convert power between different formats, such as alternating current (AC) to direct current (DC). This chapter focuses on three types of converters: rectifiers, inverters, and harmonic filters. These subjects will help you build your understanding, as this book unfolds, of how wind turbines convert wind energy to electricity.

Chapter Topics

This chapter covers the following topics and concepts:

- Basic models and methods of wind power forecasting
- Electronic devices, focusing on relays as the basic building blocks of more complex types of electronic devices, such as contactors, controllers, and switches
- Power converter devices, including rectifiers, inverters, and harmonic filters

Chapter Goals

When you complete this chapter, you will be able to:

- Discuss the importance of and examine the basic methods by which wind power is forecast
- Explain the workings of relays, focusing on electromechanical relays
- Compare and contrast the functionality and design of contactors, controllers, and switches
- Describe several reasons for use of power converters in wind energy technology
- Define particular traits of rectifiers, inverters, and harmonic filters

Wind Power Forecasting

Historically, wind has not been a very predictable phenomenon. This is especially true for predictions made more than a couple of hours in advance. Yet researchers have long been working to better their wind forecasting methods.

Wind forecasting can be of great use in wind energy generation, for several reasons. For example, if operators know with certainty that wind speeds will exceed turbine capabilities, they can shut turbines down before damage occurs. Determining how best to integrate electricity into the grid depends on demand and supply at any given time. Turbine scheduling and dispatch is better when wind forecasts are more accurate.

Weather, including wind, can be hard to predict. Thus, most wind prediction efforts have been confined to near-future forecasts. **Short-term wind forecasting** attempts to predict wind levels from 1 to 48 hours in advance. The best forecasting is based on models of current and past wind behavior. This is true for onshore and offshore developments. Forecasts are developed from statistical time series sampling and broader meteorological prediction models. Most short-term wind forecasts include the following three components:

- Numerical weather prediction model output
- Observational inputs
- Numerical forecasting model and output

All forecasting models require basic inputs such as wind farm layout and turbine power curves. Many models also consider past aggregate turbine performance. Terrain variations and roughness are micro- and meso-scale forecasting factors.

Short-term wind forecasts almost always include a margin of error (standard deviation). It is calculated for each forecast. Forecasting models also include extreme events. That includes storms or moments where no wind is expected at all.

SITE SELECTION

Monitoring specific sites and their wind speed and direction is a critical part of the process of selecting a site to place a wind turbine. Before they will approve or finance a wind energy project, both utility regulators and banks often require collection of a full year of wind data. This is typically collected by towers 30 meters high that take readings of wind speed and direction data every 15 minutes. Only once this data is in hand can the best choice of wind turbine be selected and the economic feasibility be determined.

Power Electronic Devices

The key to modern electronic circuit control lies in the opening and closing of circuits. Some circuits are open by default. Others are closed, but all controllable circuits can be closed or opened by human or machine intervention. In the following section, you will learn the basic concepts of how relays are built and controlled. Then you will build on your knowledge by learning about more complex arrangements of electronic devices.

Electromechanical Relays

Electromechanical relays are devices that both complete and interrupt circuits. They do so by making physical contact between two points that are electrically charged. When relay coils are energized, electricity flows through them and creates magnetic fields. DC units are of fixed polarity, but AC units change polarity at 120 Hz. Both AC and DC models work in the same way. One end of the armature is fixed but pivots, while the other end moves to complete the opening and closing of the circuit.

All relays are based on two electrical concepts named for their discoverer. His name was Gustav Kirchhoff.

Kirchhoff's circuit law states that the total current entering a node is equal to the current leaving the node. It is expressed mathematically as such:

$$\sum{}^I in = \sum{}^I out$$

Kirchhoff's voltage law states that the sum of voltages around a loop equals zero. It is expressed mathematically as such:

$$\sum{}^{V} Loop = 0$$

Relays and Contactors

Relays are very simple devices. They contain four main components: armatures, electromagnets (also called *coils*), springs, and contacts. In **FIGURE 2-1**, you are shown two complete circuits. Refer to this figure as you read about how basic relays work:

- Open relay—The first circuit is the "bottom" one, where the on/off switch acts as a contact. You can use it to open and close this first circuit. When you close the circuit by turning the switch on, the battery powers the electromagnet. Now move your attention to the "upper" circuit formed by the battery, light

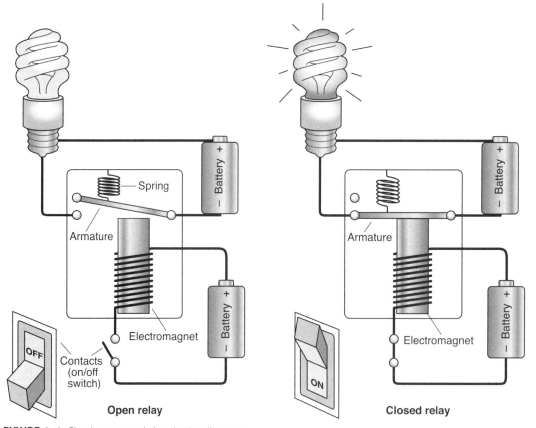

FIGURE 2-1 Simple open- and closed-relay diagrams.

Adapted from Plantier, K., & Smith, K. M. (2009). *Electromechanical principles of wind turbines for wind energy technicians.* Waco, TX: Texas State Technical College Publishing.

bulb, armature, and spring. If the armature closes, the battery will power the light bulb, but usually the spring holds the armature in the open position.

▪ **Closed relay**—When you close the bottom circuit and power the electromagnet, it attracts the armature toward it. That moves against the force of the spring and the armature moves into the closed position. The armature has closed the relay and the battery powers the light bulb.

When you open the bottom relay (turn the switch off), the electromagnet stops working. That allows the spring to pull it back into the closed position and the light bulb stops emitting light. Read on to find out more about different relay types and how relays can be combined to form controllers.

Relay Types

There are many different types of relays. Some of the different categories are time-delay relays, protective relays, and solid-state relays. Each of these categories contains different types of relays as well. Right now you will begin by focusing your attention on some **time–delay relays**.

Time-delay relays stay on for a certain amount of time once they are activated. They are made using adjustable timer circuits. This controls the actual relay. Some time-delay relays are made with specialized "shock absorbers." They can buffer armature action when the electromagnet is activated, deactivated, or both. These relays are categorized by their "normal" states, either open or closed. They are further subdivided by whether their opening or closing is time controlled:

▪ *NOTC*—This stands for **normally open, timed closed relays** (also called "normally open, on-delay" relays). Usually this type of relay is open when the coil is de-energized. When the coil has been powered for a specific amount of time, the armature closes the circuit. Refer to **FIGURE 2-2**.

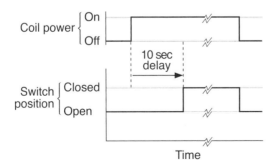

FIGURE 2-2 NOTC relay diagram.

Adapted from Plantier, K., & Smith, K. M. (2009). *Electromechanical principles of wind turbines for wind energy technicians.* Waco, TX: Texas State Technical College Publishing.

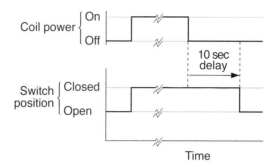

FIGURE 2–3 NOTO relay diagram.

Adapted from Plantier, K., & Smith, K. M. (2009). *Electromechanical principles of wind turbines for wind energy technicians.* Waco, TX: Texas State Technical College Publishing.

In this diagram, you can see that the circuit is open by default. After the coil has been powered for 10 seconds, the armature closes. However, the very moment the coil loses power, the armature opens again.

- *NOTO*—This stands for **normally open, timed open relays** (also called "normally open, off-delay" relays). This type of relay is also open when the coil is de-energized. However, when the coil is energized in this type of relay, the armature closes the circuit immediately. Refer to **FIGURE 2-3**. In this diagram, you can see that the circuit is open by default. As soon as the coil is powered, the armature closes the circuit. However, when the coil is de-energized, the circuit stays closed for 10 seconds. *Then* the armature goes back to its default position of open.

- *NCTO*—This stands for **normally closed, timed open relays** (also called "normally closed, on-delay" relays). This type of relay is closed when the coil is de-energized. When the coil is powered for a specific amount of time, the armature opens the circuit. As soon as the coil loses power, the circuit recloses. Refer to **FIGURE 2-4**. It represents an NCTO relay with a 10-second opening delay.

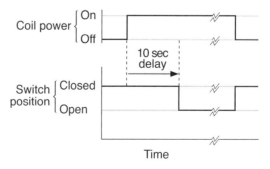

FIGURE 2–4 NCTO relay diagram.

Adapted from Plantier, K., & Smith, K. M. (2009). *Electromechanical principles of wind turbines for wind energy technicians.* Waco, TX: Texas State Technical College Publishing.

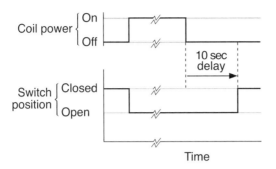

FIGURE 2-5 NCTC relay diagram.

Adapted from Plantier, K., & Smith, K. M. (2009). *Electromechanical principles of wind turbines for wind energy technicians.* Waco, TX: Texas State Technical College Publishing.

- *NCTC*—This stands for **normally closed, timed closed relays** (also called "normally closed, off-delay" relays). This type of relay is also closed when the coil is de-energized. As soon as the coil is energized, the circuit opens. However, when the coil is depowered, the reclosing of the circuit is delayed. Refer to **FIGURE 2-5**. It represents an NCTC relay with a 10-second closing delay.

Protective Relays

Protective relays shut down electrical systems or certain components under varying circumstances. Usually they consist of short circuits or abnormal currents that can interfere with or damage electrical equipment. Have you ever seen or used circuit breakers? They are a form of protective relay.

Protective relays must be placed so that each generator, bus, transformer, and other key electronic system can be completely closed off from the larger circuit. They must be able to momentarily carry the complete current of a short circuit. Then they must interrupt that current without delay once it is registered as problematic. If protective relays are too expensive for a given application, fuses can take their place.

Solid-State Relays

One problem with traditional relays is that their moving parts can wear out or fail. If the moving parts stop working, relays are useless. **Solid-state relays (SSRs)** contain no moving parts. They are classified by their input circuits' properties. There are two types of SSRs on wind turbines:

- Photo-coupled SSR—This SSR actuates via a low-voltage light-emitting diode (LED) signal. The signal is optically isolated from the rest of the circuit. A diode receives the signal and turns on a silicon-controlled rectifier called a thyristor. The thyristor switch opens and closes the circuit.

- Transformer-coupled SSR—In this version of an SSR, a combined voltage level activates the thyristor. The combined voltage consists of a control signal added to the primary voltage of a small transformer. The control signal can run via a DC/AC converter or directly via AC.

Each type of SSR has its advantages and disadvantages. However, those are outside the scope of this chapter. If you are interested in finding out more about SSRs, research them in your local library or on the Internet.

Types of AC Controllers

Relays that switch large currents into their system are called **contactors**. Contactors usually contain many switches and most are the "normally open" type. Large groups of contactors are called controllers. Controllers usually control electric motors such that when the controller coils are energized, they cut power to the motor. There are three types of AC controllers on wind turbines:

- *LVP*—This means **low-voltage protection**. These controllers depower motors in low-voltage conditions. They also keep the motors from restarting when normal voltage returns.
- *LVR*—This means **low-voltage release**. These controllers also depower motors in low-voltage situations, but unlike LVPs, LVRs restart the motor when normal voltage resumes. Small and/or important loads usually call for LVRs.
- *LVRE*—This means **low-voltage release effect**. This is a manual controller that holds motors constantly at full voltage. These are used on small, vital loads as well.

Switches

Switches can function alongside or in place of relays, contactors, and controllers. There are two main categories of switches. The first is proximity switches and the second is limiter switches. **Proximity switches** open or close circuits when the circuits get within a specific distance from another object. Here are the four common types of proximity switches:

- Capacitive proximity switches—These switches have a sensor with a metal plate attached to a radio frequency oscillator. If another object approaches the sensor, the radio frequency changes and the sensor signals the switch to open or close the circuit as necessary.
- Inductive proximity switches—These switches contain electrified coils within them. The coil measures current magnetically. If another object comes close and the current increases, the switch is signaled to open or close the circuit.
- Acoustic proximity switches—These switches measure distance via sound waves. If an object comes too close and the sound waves are detected as bouncing back too rapidly, the switch is activated to open or close the current.

- **Infrared proximity switches**—These switches function like acoustic proximity switches with one main difference. The switch opens or closes the circuit based on distance measured in infrared light reflections.

Limiter switches are most similar to a light switch, but again they can close *or* open circuits. Their main function is via a visible device connected to an electromagnetic actuator. Their structures resemble the diagram in Figure 2-2. These switches are relatively rugged, easy to see, easy to install, and very reliable.

Power Electronic Converters

Power converters convert electricity from one form to another. For example:

- AC → DC
- DC → AC
- Voltage *n* → Voltage *i*
- Frequency *x* → Frequency *y*

As technology develops and prices fall, more and more wind energy applications are available for power converters. Converters are power electronic devices made of a control system turning switches (or valves) on and off. Key electronic elements include the following:

- **Diodes**—These act as one-way valves.
- **Silicon-controlled rectifiers (SCRs)**—These are also called thyristors. They act as diodes. An external pulse turns on SCRs at the gate. The voltage across them turns them off.
- **Gate turn off thyristors (GTOs)**—These are SCRs that may be turned off and on.
- **Power transistors**—These function similarly to a GTO, but with simpler firing circuitry. Also, to stay on they require the gate signal to be applied continuously.

Review **FIGURE 2-6** before reading about the different types of power converters.

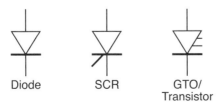

Diode   SCR   GTO/
                            Transistor

FIGURE 2-6 Schematic converter circuit elements in this chapter.

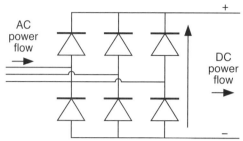

FIGURE 2-7 Simple rectifier. Input: Three-phase AC. Output: Fluctuating DC.

Adapted from Manwell, J. F., McGowan, J. G., & Rogers, A. L. (2009). *Wind energy explained: Theory, design, and application.* West Sussex, England: John Wiley & Sons, Ltd.

Rectifiers and converters are the most common power converters currently in use in the wind energy field. The following sections discuss them further; be sure to understand the importance of harmonic filters as they relate to rectifiers and inverters.

Rectifiers

Rectifiers convert AC power into DC power and they may form part of a variable-speed system or a battery-charging system. Simple rectifiers use diode bridge circuits to convert AC to fluctuating DC. See **FIGURE 2-7** for an example where the input is three-phase AC being converted to DC.

There is another type of rectifier called a controlled rectifier. They allow control over the output voltage. SCRs replace diodes in controlled rectifiers **FIGURE 2-8**. The SCRs remain off for part of the cycle, and then they are turned on. The SCR thus produces an average voltage equal to the cosine of the firing delay angle.

Inverters

Inverters convert DC to AC. Electronic inverters are made of circuit elements and control circuitry. The circuit elements switch high currents and the circuitry coordinates the switching. Inverters connected to an AC grid and taking their switching

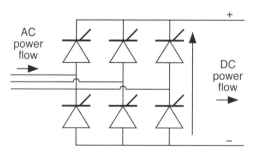

FIGURE 2-8 Controlled rectifier. Input: Three-phase AC. Output: Controlled voltage DC.

Adapted from Manwell, J. F., McGowan, J. G., & Rogers, A. L. (2009). *Wind energy explained: Theory, design, and application.* West Sussex, England: John Wiley & Sons, Ltd.

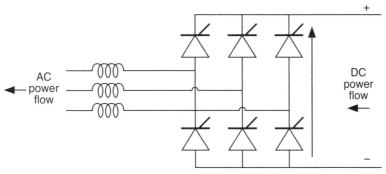

FIGURE 2-9 Line-commutated inverter. Input: DC. Output: Three-phase AC.

Adapted from Manwell, J. F., McGowan, J. G., & Rogers, A. L. (2009). *Wind energy explained: Theory, design, and application.* West Sussex, England: John Wiley & Sons, Ltd.

signal from it are line-commutated inverters **FIGURE 2-9**. The circuitry is similar to that of a controlled rectifier, but there are differences. One is that the timing of the circuit element switches is controlled externally. The other is that the current flows from DC to three-phase AC.

Inverters that are not dependent on grid connections are called self-commutated inverters. The circuitry itself can vary in design, but there are two main categories of design for it. The first category is voltage source inverters, and the second is current source inverters. Voltage source inverters are the most common type in wind energy, and they run from a constant voltage DC power supply. See **FIGURE 2-10** for an example of volt source inverter circuitry. Current source inverters hold the source's DC current constant, no matter the load. They are used in high power-factor load situations, and their efficiencies are around 96 percent. The control circuitry is complex in current source inverters.

Harmonic Filters

Harmonic filters filter **harmonic distortion**. That is the result of waveform changes caused principally by inverters, motor drives, electronic appliances, light

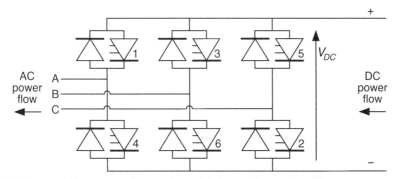

FIGURE 2-10 Voltage source inverter. Input: DC. Output: Three-phase AC.

Adapted from Manwell, J. F., McGowan, J. G., & Rogers, A. L. (2009). *Wind energy explained: Theory, design, and application.* West Sussex, England: John Wiley & Sons, Ltd.

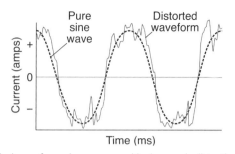

FIGURE 2-11 Typical waveform changes caused by harmonic distortion.

Adapted from Manwell, J. F., McGowan, J. G., and Rogers, A. L. (2009). *Wind Energy Explained: Theory, Design, and Application.* West Sussex, England: John Wiley & Sons Ltd.

dimmers, fluorescent light ballasts, and personal computers. Harmonic distortion can cause premature failure of the winding insulation because it overheats transformers and motor windings. Heating caused by resistance in the windings and eddy currents in the core is a function of the square of the current. So you can imagine that small increases in current have potentially large harmonic distortion side effects. See **FIGURE 2-11** for an example of harmonic distortion.

Harmonic filters are necessary to minimize the effects of harmonic distortion. These work by improving the waveform closer to a pure sine wave, and they reduce harmonics' side effects described previously. Look at **FIGURE 2-12**. It is a

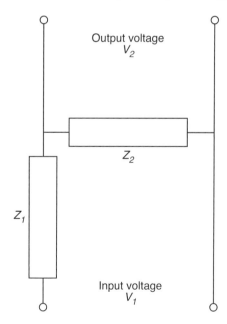

FIGURE 2-12 Simple harmonics filter. V_1: Input voltage. V_2: Output voltage. Z_1: Series impedance. Z_2: Parallel impedance.

Adapted from Manwell, J. F., McGowan, J. G., and Rogers, A. L. (2009). *Wind Energy Explained: Theory, Design, and Application.* West Sussex, England: John Wiley & Sons Ltd.

representation of a simple AC voltage filter. The input voltage is V_1 and the output voltage is V_2. Ideally, the voltage filter should result in very little change of the input voltage and a large reduction in harmonics. Did you also notice the series and parallel impedances (Z_1 and Z_2)?

SUPPORTING WIND TURBINE MANUFACTURING

Have you ever heard of the US Department of Energy's Wind and Water Power Program? This program aims to lower costs, improve performance, and accelerate deployment of wind and water power technologies in the United States. The federal government has an interest in promoting these technologies. This is because more use of the nation's wind and water resources for electricity production helps stabilize energy costs. It also enhances energy security and improves the environment. Partnering with the Department of Energy's national laboratories, this program aids wind and water power research and development. It does this through competitively selected, cost-shared research and development projects with private wind energy companies. The Wind and Water Program also forms partnerships with other federal and state stakeholder groups to achieve its mission.

The Wind and Water Power Program works with US manufacturers to develop advanced blade designs and manufacturing techniques. It also works to increase automation, which enables wind turbines to capture more energy. Automation can also help blade manufacturers increase their production capacities. For example, the program's blade manufacturing research initiatives include the following:

- More efficient turbine structures (this includes blade designs that combine structure with aerodynamics)
- Adaptive structures, like passive bend-twist coupling
- Design details to minimize stress concentrations
- More economical blade root attachment equipment

During the past few years, the program has worked with Knight and Carver, a blade production company. The work is in development for an innovative wind turbine blade design. It is the first of its kind to be produced at a utility-grade size. The Sweep Twist Adaptive Rotor (STAR) blade is 89 feet long and made of fiberglass and epoxy resin. Its revolutionary shape curves near the trailing edge. This reduces fatigue loads and increases wind energy production.

To learn more about how the federal government is investing in America's future through support of wind and water power technologies, visit the US Department of Energy's Wind and Water Power Program home page at *http://www1.eere.energy.gov/windandhydro/*.

CHAPTER SUMMARY

In this chapter, you read about the general principles behind modern power electronics and converter systems. The chapter opened with a short discussion of wind forecasting. You read about the methods behind wind forecasts. You also covered information about what wind forecast models are based on.

Next, you moved on to cover electronic devices. You read about electromechanical relays as a basis for the formation of other, more complex electronic devices. These include contactors, controllers, and switches. You also read about power converters, the devices that convert power between different formats, such as AC to DC. In this chapter, you focused on three types of converters: rectifiers, inverters, and harmonic filters.

CHAPTER KEY CONCEPTS AND TERMS

Contactors
Electromechanical relays
Harmonic distortion
Harmonic filters
Inverters
Kirchhoff's circuit law
Kirchhoff's voltage law
Limiter switches
Low-voltage protection (LVP)
Low-voltage release (LVR)
Low-voltage release effect (LVRE)
Normally closed, timed closed
   (NCTC) relays

Normally closed, timed open
   (NCTO) relays
Normally open, timed closed
   (NOTC) relays
Normally open, timed open
   (NOTO) relays
Power converters
Protective relays
Proximity switches
Rectifiers
Short-term wind forecasting
Solid-state relays (SSRs)
Time-delay relays

CHAPTER ASSESSMENT: MODERN POWER ELECTRONICS AND CONVERTER SYSTEMS

1. Wind power forecasting focuses mainly on costs saved by eliminating excess personnel during storms or periods with no wind.
   - ❏ A. True
   - ❏ B. False

2. Short-term wind forecasting attempts to predict wind levels from _____ in the future at any given moment.

3. Which of the following items is not normally a wind power forecast input?
   - ❑ **A.** Wind farm layout
   - ❑ **B.** Turbine power curves
   - ❑ **C.** Terrain variation and roughness
   - ❑ **D.** Standard deviation

4. Electromechanical relays are always used to interrupt circuits.
   - ❑ **A.** True
   - ❑ **B.** False

5. Kirchhoff's circuit law states which of the following?
   - ❑ **A.** The total current entering a node is equal to the current leaving the node.
   - ❑ **B.** The total current entering a node is unequal to the current leaving the node.
   - ❑ **C.** The coefficient of voltages around a loop equals zero.
   - ❑ **D.** The sum of voltages around a loop equals zero.

6. Which of the following statements about time-delay relays is false?
   - ❑ **A.** They stay on for a certain amount of time once they are activated.
   - ❑ **B.** They are made using adjustable timer circuits.
   - ❑ **C.** They always contain specialized "shock absorbers."
   - ❑ **D.** They are categorized by their "normal" states, either open or closed.

7. Normally closed, on-delay relays are also known as:
   - ❑ **A.** NCTO relays.
   - ❑ **B.** NCTC relays.
   - ❑ **C.** NOTO relays.
   - ❑ **D.** NOTC relays.

8. Circuit breakers are an example of a protective relay.
   - ❑ **A.** True
   - ❑ **B.** False

9. Which of the following switch types uses electrified coils as the main functional equipment?
   - ❑ **A.** Capacitive proximity switch
   - ❑ **B.** Inductive proximity switch
   - ❑ **C.** Acoustic proximity switch
   - ❑ **D.** Infrared proximity switch

10. Infrared proximity switches open or close circuits based on distance measured in ultraviolet light reflections.
    - ❑ **A.** True
    - ❑ **B.** False

11. Simple rectifiers use _____ to convert AC to fluctuating DC.

12. Inverters convert _____.

13. Which of the following statements about control source inverters is true?

❑ **A.** Their control circuitry is simple.

❑ **B.** Their efficiencies are in the 75 percent range.

❑ **C.** They hold the source's AC current constant, no matter the load.

❑ **D.** They are used in high power-factor load situations.

14. Inverters that do not depend on grid connections are called _____.

15. Harmonic filters work by improving the waveform closer to a(n) _____.

Fixed-Speed Induction Generators

IN THIS CHAPTER, you will begin your study of fixed-speed induction generators. This kind of generator is one of the most common types of generators in use on wind turbines today. This chapter opens with an overview of induction generators, differentiating the two main types: squirrel cage generators and wound-rotor generators. You will also read about why induction generators are so common in wind energy.

Next, the chapter moves on to review the main characteristics of induction generators. These revolve around the main functionalities of induction generators. You will read about the concept of slip and you will find out why some electricity produced by generators never makes it to the grid. Next, you will read about the main generator design concerns, and, finally, you will cover information about typical generator operations.

Chapter Topics

This chapter covers the following topics and concepts:

- Fixed-speed induction generator overview
- Fixed-speed induction generator characteristics
- Fixed-speed induction generator design
- Fixed-speed induction generator control

Chapter Goals

When you complete this chapter, you will be able to:

- Describe the two main types of fixed-speed induction generators
- Discuss the reasons that induction generators are the most common generators in wind energy
- Diagram the equivalent current of a typical induction generator
- Calculate the slip of an induction generator, given the necessary values
- Recall the main reasons for energy loss in induction generators
- Examine the main induction generator design concepts
- Compare and contrast the two main methods of induction generator startup and control
- Discuss the reason why starting across the line is not a desirable generator startup method, and explain what method is preferable for that action

Fixed-Speed Induction Generator Overview

You know that generators are vital to wind energy production. Fixed-speed induction generators (FSIG) are one of several different types of generators used in wind turbines, but they are the most common generators found on modern wind turbines. Induction generators are also known as asynchronous generators.

There are two types of induction generators. The first type, shown in **FIGURE 3-1**, is called a **squirrel cage generator**. This is because the rotor does not contain coil windings.

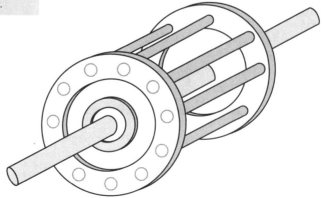

FIGURE 3-1 Example of a squirrel cage generator.

FIGURE 3-2 Example of a wound-rotor generator.

Rather, it is made of two rings connected by conducting bars. The bars conduct electric currents and are covered by a laminate to protect them. Thus, the rotor's appearance is that of a "squirrel cage."

The second type of induction generator, shown in **FIGURE 3-2**, is called a **wound-rotor generator**. That is because rather than conducting bars, the generator has copper coil windings to transmit electrical currents.

FSIGs are popular for three reasons:

- Simple, durable construction
- Comparatively economical costs
- Ease of grid connections

Fixed-Speed Induction Generator Characteristics

Induction generators function in the following manner:

1. The stator's windings are placed at 120-degree intervals around the **stator**. This produces a rotating magnetic field in the stator. The stator is a round, static metal casing that contains copper windings or conducting rods. It functions as a canister that fixes the rotor in a parallel axis. This allows the rotor to rotate inside the round space via bearings or slip rings.
2. This rotating field spins at synchronous speed.
3. The **rotor** spins at a slightly different speed than the stator's rotating field. The rotor is a conductive metal cylinder that spins inside the stator. It is connected to the turbine rotor via the drive train. The rotor's spinning action inside the stator allows for the conversion from mechanical energy delivered by the turbine rotor into electricity fed into the grid. Electricity is derived

from the relative motion between the rotor and the stator's magnetic field.

4. The rotating stator magnetic field induces electromagnetic currents in the rotor.

5. Interaction between the stator field and the induced rotor field causes elevated voltage in the terminals. This voltage flows out of the terminals into conducting cables that eventually connect to the grid.

A key characteristic of induction generators is **slip**. Slip is defined as the ratio between synchronous speed and rotor operating speed. The formula for determining slip is:

$$s = \frac{n_s - n}{n_s}$$

In the preceding equation, s = slip, n_s = synchronous speed, and n = rotor operating speed. When the values of this equation yield a positive result, the machine functions as a motor. When the slip results are negative, the machine functions as a generator.

The prevalent characteristics of induction generators are shown in **FIGURE 3-3**, which is a schematic diagram. It represents the equivalent current of induction generators.

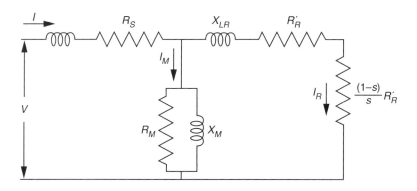

V = Terminal voltage
I = Stator current
I_M = Magnetizing current
I_R = Rotor current
X_{LS} = Stator leakage inductive reactance

R_S = Stator resistance
X'_{LR} = Rotor leakage inductive reactance referred to the stator
R'_R = Rotor resistance referred to the stator
X_M = Magnetizing reactance
R_M = Resistance in parallel with mutual inductance

FIGURE 3-3 Schematic of induction generator equivalent current.

Adapted from Manwell, J. F., McGowan, J. G., & Rogers, A. L. (2009). *Wind energy explained: Theory, design, and application*. West Sussex, UK: John Wiley & Sons, Ltd.

Not all electricity generated by induction generators is useful. This is because there are some power losses after generation. The main losses are caused by the following:

- Mechanical issues such as wind friction—These losses are caused by rotor drag from air friction.
- Rotor and stator resistive and magnetic losses—These losses are caused principally by bearing electricity absorption.

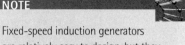

NOTE

Fixed-speed induction generators are relatively easy to design, but they are also very large and heavy due to all the copper windings. The mass of copper involved also makes them rather costly. A typical generator can be as large as 3 meters in diameter and 3 meters long, weighing in at over four tons for large modern 2.5-megawatt generators.

Fixed-Speed Induction Generator Design

The design considerations of induction generators are similar to those of all generators. Stator housings are generally steel and come in standardized sizes. Turbine manufacturers set the standards. That is because the generator has to fit into the nacelle housing with other turbine components.

Windings are made of copper cables laid into grooves in the stator and rotor (in wound-rotor designs). The cables are insulated to ensure protection from the environment and to provide stability.

The generator's exterior is designed with the interior in mind. In other words, it protects the interior from dew formation, dust particles, insects, and the like. There are two common designs. One is the "open drip-proof" design and the other is the "totally enclosed, fan-cooled" (TEFC) design.

The open drip-proof design is the most common because usually the nacelle housing is sufficient to protect inner components from environmental dangers. However, recent studies show that TEFC designs do provide additional protection from damage. Thus, TEFC generator exteriors are becoming more common in newer generator designs. Refer to **FIGURE 3-4** for a cutaway diagram showing a typical fixed-speed induction generator design.

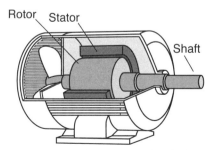

FIGURE 3-4 Cutaway of a typical induction generator showing internal components.
Adapted from Manwell, J. F., McGowan, J. G., and Rogers, A. L. (2009). *Wind Energy Explained: Theory, Design, and Application.* West Sussex, England: John Wiley & Sons Ltd.

Fixed-Speed Induction Generator Control

In most wind turbine applications, there are two methods of bringing induction generators online. The first method is to use the turbine rotor to speed up the generator rotor. Then the generator is connected to the grid and feeds electricity into it. If this method is used, the turbine must be a self-starting model. These models usually employ pitch-controlled blades. The generator is connected to the grid when it is spinning as close to synchronous speed as possible.

The second method is the inverse of the first. The generator is first connected to the grid. Then it functions like a motor to bring the turbine rotor up to operating speed. In general, across-the-line starting is not the preferred method for this starting technique. Rather, some form of current reduction helps maintain the generator's life span and reliability rates. This is because across-the-line starting can wear down windings over time.

CAREERS IN WIND ENERGY

You want to have a career in wind energy so you are studying hard, but how else can you begin to prepare for your future career? Nowadays, many websites where people can share similar experiences are available. Wind energy is no different. These are online sites and forums that give users the chance to participate in discussions about aspects of this exciting field. Users log on and find out where the newest and best-paying jobs are. They can also discuss various aspects of the job, such as safety issues, new technological advances, and green energy industry initiatives.

These online communities connect people in ways that were not possible as recently as even a few years ago. You can now join one or more and begin to connect with people who share your goals. You can also connect with current wind energy technicians, supervisors, executives, and contractors to find out more about the variety of non-technician jobs that will be available to you when you complete your training. Often, these sites also alert users to upcoming conferences and seminars on wind energy. These events can also be beneficial to attend. Attendance allows you to expand your network of personal connections.

The following is a list of some of the most common social networking sites for people interested in wind energy. Have a look at some of them and dive in. You can also do some online searching of your own. That will result in many other sites and forums for you to review and start participating in. You will find your career much more rewarding with more possibilities for advancement if you take advantage of the benefits that being connected to others in the same field has to offer.

Careers in Wind
www.careersinwind.com
The Environment Site
www.theenvironmentsite.org/forum/wind-energy-forum

CAREERS IN WIND ENERGY (Continued)

LinkedIn Wind Energy Forum
www.linkedin.com/groups?home=&gid=104093

Nature 2 Energy
http://nature2energy.com/f4/

Wind Energy Maintenance Forum
www.windenergymaintenanceforum.com/redHome.aspx?region=northamerica&lang type=1033

CHAPTER SUMMARY

In this chapter, you have studied fixed-speed induction generators. These are some of the most common types of generators in use on wind turbines today. The chapter opened with an overview of induction generators, differentiating the two main types: squirrel cage generators and wound-rotor generators. You also discovered why induction generators are so common in wind energy.

You then moved on to review the main characteristics of induction generators, which revolve around the main functionalities of induction generators. You read about the concept of slip and you focused on why some electricity produced by generators never makes it to the grid. Next, you read about the main generator design concerns, and, finally, you covered information about typical generator operations.

CHAPTER KEY CONCEPTS AND TERMS

Rotor
Slip
Squirrel cage generator
Stator
Wound-rotor generator

CHAPTER ASSESSMENT: FIXED-SPEED INDUCTION GENERATORS

1. Wound-rotor generators contain laminate-covered metal conducting rods.
   - ❑ A. True
   - ❑ B. False

2. Induction generators are also known as _____.

3. Which of the following items is *not* a reason for the popularity of fixed-speed induction generators?
   - ❑ A. Simple, durable construction
   - ❑ B. Comparatively economical costs
   - ❑ C. Ease of maintenance operations
   - ❑ D. Ease of grid connections

4. Stator windings are typically placed at 130-degree intervals.
   - ❑ A. True
   - ❑ B. False

5. Interactions between the stator field and the induced rotor field cause _____ in the terminals.

6. When slip results are positive, the induction generator functions as a motor.
   - ❑ **A.** True
   - ❑ **B.** False

7. The copper winding cables are _____ to ensure protection from the environment and to provide stability.

8. TEFC generator exteriors are the most common current generator exteriors in use today.
   - ❑ **A.** True
   - ❑ **B.** False

9. Across-the-line starting can wear down _____ over time.

10. Induction generators can be used as motors to start the wind turbine's rotor moving only if the turbine is a self-starting model.
    - ❑ **A.** True
    - ❑ **B.** False

Synchronous Generators for Wind Turbines

THIS CHAPTER BEGINS with a review of the basic workings of generators and motors. From there, you will delve into the particularities of synchronous generators, including their construction and unique characteristics. You will also read about the unique traits of synchronous generator functionality. After studying the basics of synchronous generators, you will review their design and learn about some variables in how synchronous generators can be constructed. Last, the chapter concludes with a discussion about synchronous generator control. This chapter discusses several standard control mechanisms for regulating voltage related to network load as well as generator protection from surges. When you complete this chapter, you will be able to enumerate these items and describe their functions.

Chapter Topics

This chapter covers the following topics and concepts:

- Synchronous generator functionality overview
- Synchronous generator characteristics, with a focus on winding pairs and the interaction between stator and rotor magnetic fields
- Synchronous generator design, including coil replacements and different types and uses of damper windings
- Synchronous generator control mechanisms, including several excitation control mechanisms as well as prime mover control components

Chapter Goals

When you complete this chapter, you will be able to:

- Describe the basic theory behind motor and generator functionality
- Compare and contrast the similarities in functionality between motors and generators
- Describe the particular basic traits of salient-pole and wound-rotor synchronous generators
- Diagram the common stator and rotor winding patterns for both types of synchronous generators
- Describe the interactions between the rotor and stator magnetic fields, focusing on synchronous speed in the field rotation
- Describe the uses of coils and damper windings as two possible add-on components of synchronous generators
- Discuss the four components of excitation control and the importance of prime mover controls with respect to synchronous generator terminal voltage

NOTE

Presently, synchronous generators are the most commonly used, but as electronics control system technology improves, this may change. Much research is currently going into making generators both lighter and smaller for wind turbine applications.

Synchronous Generator Overview

Motors and generators generally perform opposite functions. Motors convert electric energy into mechanical energy, so generators convert mechanical energy into electric energy. Both are generically called electrical machines because they can perform either task. In this chapter, you will focus on generators. You will not read specifically about motors, but in general you can think of a motor as a "reverse" generator in terms of how it works.

FIGURE 4-1 illustrates a basic generator. Imagine the shaft, or armature, spinning due to mechanical energy coming from the main shaft and gearbox. The spinning creates a voltage, as Faraday's law says that it should. If the cable carrying electricity down the tower is connected by slip rings, there is one connected to the armature coil and one connected to the other coil. The slip rings contain brushes that allow the current to pass from one ring to the other. The voltage generated by the armature's spinning depends on its position in the magnetic field. It varies sinusoidally when spinning at a constant rate. Thus the generator is operating in alternating current (AC) mode.

Commutators can replace slip rings. A simple commutator can be made of two 180-degree segments. The brushes contact one segment at a time. In each revolution, brush-segment pairings reverse one time. Thus the induced voltage would be made of half-sine waves that all have the same sign. Commutators are used in most common direct current (DC) motors.

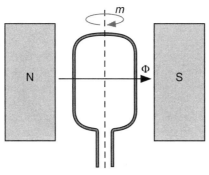

m = mechanical energy   Φ = magnetic flux
N = north magnetic pole   S = south magnetic pole

FIGURE 4-1 Example of a simple generator.

Adapted from Manwell, J. F., McGowan, J. G., and Rogers, A. L. (2009). *Wind Energy Explained: Theory, Design, and Application.* West Sussex, England: John Wiley & Sons Ltd.

Most generators function like the simple one you just read about. See **FIGURE 4-2** for an example of typical asynchronous generator construction. Here are some key differences:

- **Permanent magnets**—Most generators do not have permanent magnets. The fields are usually produced electronically. Traditionally, permanent magnets have had demagnetization problems and high costs. This is changing, but commercial wind turbine generators still do not usually contain permanent magnets.
- **Fields and armature**—The fields are usually on the rotor, which rotates. The stationary part usually houses the armature.
- **Armature field**—The armature produces its own magnetic field, and it can interact with the rotor's fields. Both fields affect generators' operating efficiencies.

Induction (asynchronous) generators and synchronous generators are the two most common types. DC generators are a less common third type of generator used in wind turbines.

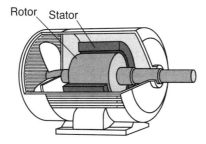

Rotor   Stator

FIGURE 4-2 A typical three-phase asynchronous (induction) generator.

Adapted from Manwell, J. F., McGowan, J. G., and Rogers, A. L. (2009). *Wind Energy Explained: Theory, Design, and Application.* West Sussex, England: John Wiley & Sons Ltd.

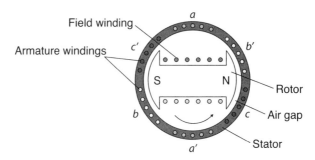

FIGURE 4-3 Cross-sectional view of a three-phase synchronous generator.

Adapted from Anaya-Lara, O., Jenkins, N., Ekanayake, J., Cartwright, P., & Hughes, M. (2009). *Wind energy generation: Modeling and control.* West Sussex, England: John Wiley & Sons, Ltd.

Like asynchronous generators, synchronous generators contain two main parts. The first part is the **field winding**, which is located on the rotor. The second part is the **armature winding**, located on the stator. The armature is a three-phase winding, as shown in FIGURE 4-3.

A direct current flows through the field winding. The magnetic field rotates as the rotor turns. Low-speed generator rotors contain concentrated winding and a nonuniform air gap. This kind is called a salient-pole generator. High-speed generator rotors have distributed winding and a uniform air gap. This kind of generator is called a wound-rotor or *cylindrical pole* generator.

Both types of synchronous generators produce sinusoidal magnetic fields in their air gaps. Salient-pole generators' poles are shaped to create a sinusoidal air-gap flux. Rotor windings in wound-rotor generators are distributed over two-thirds of the rotor surface. Their produced flux aggregates into a sinusoidal shape. Moreover, stator winding placement helps to create voltage with sinusoidal waveforms. In this chapter, you will focus on the workings and modeling of salient-pole generators. However, functionality for both types of synchronous generators is effectively the same.

Synchronous Generator Characteristics

Understanding synchronous generators revolves around windings and the magnetic field of the air gap. Three pairs of stator windings with axes are spaced 120 degrees apart. When the rotor turns, its magnetic field spins in the air gap at synchronous speed. The spinning field cuts the stator's three voltages. Currents are then induced in the three pairs of stator windings. If the stator windings are connected to identical loads, a three-phase current results. Of course the three-phase currents are also displaced by 120 degrees. The displaced currents will also create three magnetic fields. Thus the air gap contains a combination of magnetic fields. Both fields produced by the stator currents as well as the rotor currents are present.

You can see in FIGURE 4-4 that when $t = 0$, the a windings are at their maximum level. Conversely, the b and c windings are at their half-maximum negative currents. At t_1 the c windings reach their maximum negative, while the a and b

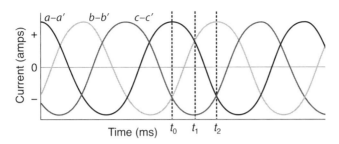

FIGURE 4-4 Diagram of three-phase currents in the stator windings.

Adapted from Anaya-Lara, O., Jenkins, N., Ekanayake, J., Cartwright, P., & Hughes, M. (2009). *Wind energy generation: Modeling and control.* West Sussex, England: John Wiley & Sons, Ltd.

currents coincide at their half-maximum positive currents. At t_2 the b windings reach their maximum current. You can see that if the timescale continued, the c currents would reach their maximum at t_4.

Continuing, at t_1 the a and b phase currents make two magnetic fields. The fields' magnitudes are proportional to the number of ampere-turns along the a and b axes. Likewise, the current in the c phase is proportional to the number of ampere-turns relative to the c axis. Given these interactions, you can deduce that the stator magnetic field at t_1 shifts by $\frac{\pi}{3}$. See **FIGURE 4-5**.

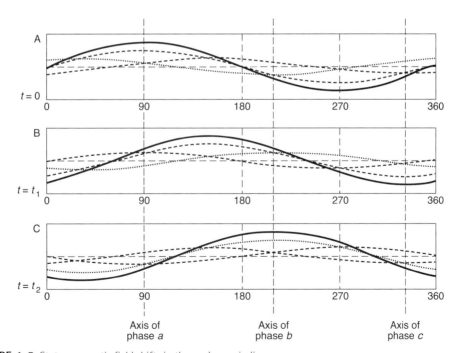

FIGURE 4-5 Stator magnetic field shifts in three-phase windings.

Data from Anaya-Lara, O., Jenkins, N., Ekanayake, J., Cartwright, P., & Hughes, M. (2009). *Wind energy generation: Modeling and control.* West Sussex, England: John Wiley & Sons, Ltd.

In Figure 4-5, you can see that within each phase, the stator's magnetic field has rotated by $\frac{\pi}{3}$. This is also known as rotating at synchronous speed.

Synchronous Generator Design

Now that you know the basic functionality of synchronous generators, you can delve into some of the design considerations and alterations that are possible.

> **NOTE**
>
> There are some new generators that use superconducting materials for their windings. They are smaller and lighter but very costly to manufacture and operate, since they require extensive cooling to maintain superconducting performance of the wires.

Synchronous generator design is similar to that of asynchronous generator design. The stator and rotor are usually constructed of a minimally conductive steel alloy. The stator and rotor windings for both salient-pole generators and wound rotor generators are most often made of copper or copper alloys.

Synchronous Generator Coil Usage

With balanced three-phase currents, a coil sometimes replaces the stator windings. This single coil is aligned with the *a* phase axis. The coil carries the same sort of current of stator windings. Its magnetic field rotates likewise at synchronous speed. This variant of generator design is most common in studying steady-state designs. In real-life applications, the coil design is rarely possible. This is because currents change in magnitude and phase in the three-phase windings. Also, the stator field vector does change with the use of the coil.

Synchronous Generator Damper Windings

Another synchronous generator design consideration pertains to damper windings. Salient-pole and wound-rotor designs both contain solid copper bars running through the rotor. The copper bars provide additional circulatory paths for damping currents.

Salient-pole generators contain the damper bars inlaid into pole faces. You can see this arrangement in **FIGURE 4-6**. The figure shows discontinuous end rings. However, some generators have end rings that link poles. This provides even more pathway for damping current flow.

The damper windings in cylindrical pole generators are a little different. In **FIGURE 4-7** you can see that the damper windings are embedded into a rotor's slot wall. The slot wall contains the field winding and the damper winding on top. These are held in place by a specialized wedge. All wound-rotor generators contain end rings that connect the dampers. However, for better visibility, these end rings are not included in Figure 4-7.

In both types of synchronous generators, damper windings function in the same way. The damper winding currents combine with the air-gap flux. This produces torque. The torque then dampens rotor fluctuations after any kind of transient disruption.

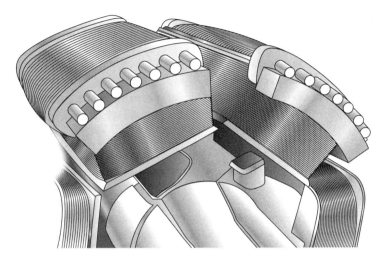

FIGURE 4-6 Inlaid current damper windings in a salient-pole generator.

Adapted from Anaya-Lara, O., Jenkins, N., Ekanayake, J., Cartwright, P., & Hughes, M. (2009). *Wind energy generation: Modeling and control.* West Sussex, England: John Wiley & Sons, Ltd.

Sometimes the grid to which a generator is connected experiences an unsymmetrical fault. This causes the air-gap flux to develop two components. The two components are **positive sequence** and **negative sequence**. Positive sequence is defined as flux resulting from currents in the direction of rotation. Negative sequence is flux resulting from currents flowing opposite to the rotational direction. The negative sequence counteracts the rotational direction of the rotor. Thus a high relative speed develops, and this produces a large torque aspect of the generator's operation.

Damper winding currents interact with negative sequence in the air-gap flux. That produces a counteracting torque. The result is that acceleration is controlled, and the generator's rate of increase is limited.

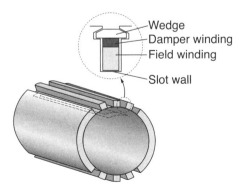

FIGURE 4-7 Embedded current damper windings in a wound-rotor generator.

Adapted from Anaya-Lara, O., Jenkins, N., Ekanayake, J., Cartwright, P., & Hughes, M. (2009). *Wind energy generation: Modeling and control.* West Sussex, England: John Wiley & Sons, Ltd.

Synchronous Generator Control

Large electrical systems consist of different components that are interconnected. The main components are generators, transmission and distribution systems, and loads. Connected loads vary continuously over time. Power systems must operate within certain parameters and maintenance of system stability is paramount. Thus generators must be individually and collectively controlled. The functional controls are divided into two main functions. The first is reactive power and voltage control. The second is active power and frequency control. These functional controls are diagrammed in **FIGURE 4-8**.

Excitation Control

Active and reactive power demands vary as conditions on the system change. When loads are heavy, the transmission system absorbs reactive power. In such cases, synchronous generators add reactive power into the grid. When loads are light, transmission systems can become overloaded. In those cases, synchronous generators must absorb reactive electricity from the system. Due to these varying generator demands, an **excitation control system** is needed. Excitation control systems provide automatic voltage regulation and they protect the generator from excess power surges. **FIGURE 4-9** is a diagram of major excitation control system components.

Regulators

Synchronous generators contain **automatic voltage regulators (AVRs)**. Automatic voltage regulators keep the stator's terminal voltage within defined limits. Sometimes increased demand causes terminal voltage to fall. When that

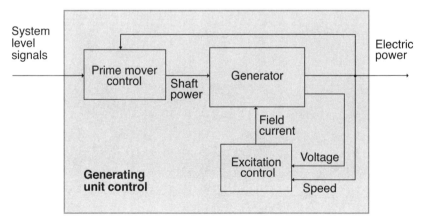

FIGURE 4-8 Schematic diagram of synchronous generator control functions.
Adapted from Anaya-Lara, O., Jenkins, N., Ekanayake, J., Cartwright, P., & Hughes, M. (2009). *Wind energy generation: Modeling and control.* West Sussex, England: John Wiley & Sons, Ltd.

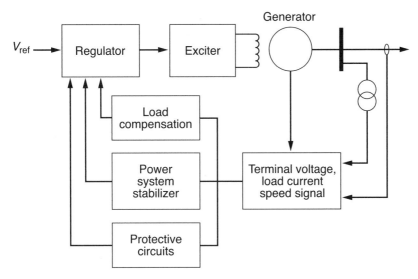

FIGURE 4–9 Schematic diagram of synchronous generator excitation control system.

Adapted from Anaya-Lara, O., Jenkins, N., Ekanayake, J., Cartwright, P., & Hughes, M. (2009). *Wind energy generation: Modeling and control.* West Sussex, England: John Wiley & Sons, Ltd.

happens, the AVR augments voltage. That is triggered via an electronic signal fed into the AVR. Terminal voltage then returns to its initial value.

Exciters

Exciters supply adjustable direct currents to a generator's field winding. They can even be generators in their own right. This is usually the case on small-scale operations. Commutation problems prevent the usage of DC generators. When that is the case, AC generators supply the field by way of rectifiers.

Also, static exciters may be made from controlled rectifiers with currents coming from generator terminals. In all of the varied setups described, the DC supply usually connects to the synchronous field via slip rings.

Avoidance of slip rings and brushes, which require a lot of maintenance, is possible. In such cases, the AC generator's field is mounted on the stator. Then the three-phase output is mounted to the rotor. Then the rectifier can be placed on the exciter-generator shaft. The end result is a direct connection from exciter to generator without the use of slip rings.

Load Compensators

AVRs usually control generator stator terminal voltage. However, additional loops can be built into AVRs. These loops allow grid voltage to be controlled from a remote point. These loops are called **load compensators**. Load compensators have adjustable resistance and reactance. The reactance simulates impedance between generator terminals and the point of voltage control. Using the impedance and measured current allows voltage drop to be computed. Then it is added to the terminal voltage.

Power System Stabilizers

Power system stabilizers (PSSs) dampen generator rotor oscillation. They do that by controlling excitation within the rotor. The most common stabilizing signals for excitation control are shaft speed, terminal frequency, and energy level.

Prime Mover Control

Prime movers provide governance for adjusting generator power outputs. This is necessary to keep the output in line with network load demand. Suppose a network's load increases. This adds torque to the generator and causes them to decelerate. The regulating prime mover's governor detects the speed decrease. Then, it boosts generator output to match the network load.

The change in power output detected by the governor depends on the governor's droop setting. For example, a 4 percent droop setting means that a 4 percent change in speed will trigger a 100 percent change in generator output.

In steady-state conditions, all generators linked to a network operate at the same frequency. That frequency dictates the speeds of individual generators' prime movers. Thus after a network load increase, network frequency will fall. This continues until the total power output changes produced in the regulating generators equal the change in network load. See **FIGURE 4-10** for a schematic diagram of prime mover control governance functionality.

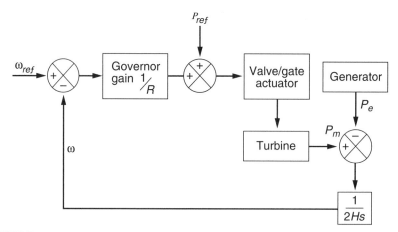

FIGURE 4-10 Schematic diagram of prime mover control governance system.
Adapted from Anaya-Lara, O., Jenkins, N., Ekanayake, J., Cartwright, P., & Hughes, M. (2009). *Wind energy generation: Modeling and control.* West Sussex, England: John Wiley & Sons, Ltd.

HOW MUCH ELECTRICITY CAN A SINGLE WIND TURBINE GENERATE?

As you know, wind turbines are large machines that convert wind's kinetic energy into electricity. Most turbine systems include, at a minimum, the following equipment:

- **Blades and rotor**—These convert the wind's kinetic energy into rotational energy, also known as torque.
- **Generator**—The generator converts torque-based energy into electricity via the mechanisms you studied in this chapter.
- **Nacelle housing**—The nacelle housing provides protection from the elements. Usually the main drive shaft, gearbox, and generator are the principal enclosed parts.
- **Tower**—The tower supports the rotor and drive train. It also provides sufficient height to ensure that blades are exposed to optimal wind levels.
- **Electronic components**—Other electronic equipment consists of cables, connectors, transformers, and grounding equipment.

Electricity generation capacity is measured in watts. You should remember the energetic level of a watt from previous courses. Watts are very small units of power measurement. Thus wind generation is measured in kilowatts (kW; 1,000 watts), megawatts (MW; 1 million watts), and gigawatts (GW; 1 billion watts).

Turbine output depends on the size of the turbine and local wind speed. Small variances in turbine size and/or wind speed can have large effects on the amount of electricity produced. This in turn can have a large effect on the price of produced energy.

You can see in the chart in **FIGURE 4-11** that since the early 1990s, turbine sizes, rating, and production have all increased by large degrees. The average rating of a commercially operative turbine in 2008 was 1.67 MW.

In 2008, wind power provided over 1.25 percent of total US electricity demand. In 2009, that number surpassed 2 percent. However, the potential for wind energy production leaves much room for growth. The US Department of Energy has a stated goal that by 2030, the United States will obtain 20 percent of its electricity from wind energy. This is because

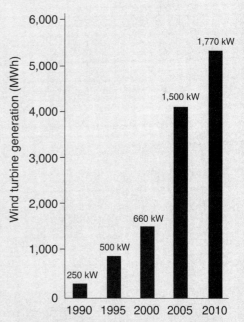

FIGURE 4-11 This chart shows the dramatic increases in wind turbine productivity from 1990 to 2010.

Data from http://awea.org/learnabout/publications/upload/AmericanWindpowerBrochure.pdf

(Continues)

HOW MUCH ELECTRICITY CAN A SINGLE WIND TURBINE GENERATE? (Continued)

wind power is inexhaustible and does not contribute to environmental pollution in the same way as other power production methods.

These amazing numbers will only grow into the future. One turbine at a time, the United States has become the world's largest gross producer of wind energy. And this industry is only just beginning to get the recognition of success it deserves. So you can count on your career in wind energy being a long and rewarding one.

CHAPTER SUMMARY

In this chapter, you started with a review of the basic workings of generators and motors. From there, you moved into the particularities of synchronous generators. This includes their construction and unique characteristics as compared with asynchronous generators. You read about the unique traits of synchronous generator functionality. After studying the basics of synchronous generators, you reviewed their design considerations and some variables in how synchronous generators can be constructed. Last, the chapter concluded with a discussion about synchronous generator control. You covered several standard control mechanisms for regulating voltage related to network load and for generator protection from network surges. You should now be able to enumerate these items and describe their functions.

CHAPTER KEY CONCEPTS AND TERMS

Armature winding

Automatic voltage regulators (AVRs)

Excitation control system

Exciters

Field winding

Load compensators

Negative sequence

Positive sequence

Power system stabilizers (PSSs)

Prime movers

CHAPTER ASSESSMENT: SYNCHRONOUS GENERATORS FOR WIND TURBINES

1. Motors and generators share the same basic structural elements and are always used as both types of electric machines.
   - ❏ **A.** True
   - ❏ **B.** False

2. Why do most modern generators *not* use permanent magnets?
   - ❏ **A.** Magnets are cheap and tend to break easily.
   - ❏ **B.** Magnets are expensive and heavier than coil windings.
   - ❏ **C.** Magnets tend to depolarize over time if exposed to the elements.
   - ❏ **D.** Magnets produce fields that are hazardous for wildlife living near wind turbines.

3. Synchronous generators have _____ sets of copper windings that create magnetic fields interacting with each other.

4. Salient-pole generators contain _____ windings and a(n) _____ air gap.

5. Wound-rotor generators contain rotor windings distributed over _____ of the rotor surface.

6. Synchronous generator stator winding pairs are spaced at _____ angles apart.

7. When generator stator fields rotate at $\frac{\pi}{3}$ within each phase, the rotation speed is called _____ speed.

8. Coils rarely replace stator windings in commercial synchronous generator operations due to which of the following reasons?
   - ❏ **A.** Currents change in magnitude and phase in the three-phase windings.
   - ❏ **B.** Currents are stable in magnitude and phase in the three-phase windings.
   - ❏ **C.** The stator field vector does not change with the use of coils.
   - ❏ **D.** The rotor field vector changes with coil usage.

9. _____ generators contain damper bars inlaid into pole faces.

10. Infrared proximity switches open or close circuits based on distance measured in ultraviolet light reflections.
    - ❏ **A.** True
    - ❏ **B.** False

11. All wound-rotor generators contain end rings that connect damper windings and provide additional circulatory paths for current damping.
    - ❏ **A.** True
    - ❏ **B.** False

12. Which of the following statements about damper windings is untrue?
    - ❏ **A.** Damper winding construction for both kinds of synchronous generators is the same.
    - ❏ **B.** Damper winding currents interact with the air-gap flux, producing torque.
    - ❏ **C.** Torque produced by the interaction between damper windings and the air-gap flux dampens current fluctuations after transient disruptions.
    - ❏ **D.** End rings connect some damper windings, but others are not connected by end rings.

13. Excitation control systems provide automatic voltage regulation and protect generators against excess electricity production.
    - ❏ **A.** True
    - ❏ **B.** False

14. Load compensators have adjustable _____ and _____.

15. The change in generator power output is detected by the prime mover control governor's _____ settings.

Doubly Fed Induction Generators

IN THIS CHAPTER, you will focus on doubly fed induction generators. They are similar to fixed-speed induction generators, but there are some important differences. You will begin by studying the basic workings of doubly fed induction generators. This includes mechanisms that permit variable-speed operations. You will also cover two methods these generators use to feed power into a grid. Then you will read about the characteristics of doubly fed induction generators. These revolve around mathematical principles. Based on these principles, this chapter covers the method for calculating doubly fed induction generator power generation. Next comes an overview of doubly fed induction generator design. Doubly fed induction generator design is mostly the same as that of fixed-speed induction generators. The last section covers two methods of power and voltage control. The first model is the most common. It is the rotor flux magnitude and angle control method. The second method is the current-mode torque control method.

Chapter Topics

This chapter covers the following topics and concepts:

- Basic doubly fed induction generator functionality, including two different methods of power delivery to the grid
- Doubly fed induction generator characteristics, focused on the calculation of power delivered from the generator to the grid
- Doubly fed induction generator design concepts
- Doubly fed induction generator control, focused on two methods of power and voltage control

Chapter Goals

When you complete this chapter, you will be able to:

- Explain the basic functionalities of doubly fed induction generators
- Describe the two methods that doubly fed induction generators use to feed power into a grid
- Describe the mathematical characteristics of doubly fed induction generator power generation
- Calculate true power generation figures given certain variables
- Discuss the design of doubly fed induction generators
- Explain the multiple roles of power converters in doubly fed induction generators
- Compare and contrast the *rotor flux magnitude and angle control* and *current-mode torque control* power and voltage control methods

TECH TIPS

Wind turbines with DFIGs are a good choice when an area has highly variable wind over the course of a year due to their ability to provide efficient power generation at high rpm.

Doubly Fed Induction Generator Overview

Doubly fed induction generators are also called DFIGs. They provide the most efficient way of converting available wind power. This is because they operate at variable speeds. This allows flexibility for wind turbine operators. This flexibility is necessary. It is related to varying wind speeds and other environmental conditions.

FIGURE 5-1 shows a typical configuration of doubly fed induction generator turbines. Doubly fed induction generators use a wound-rotor configuration.

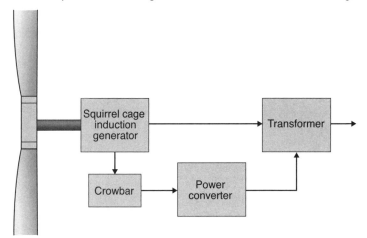

FIGURE 5-1 Typical doubly fed induction generator turbine configuration.

Adapted from Anaya-Lara, O., Jenkins, N., Ekanayake, J., Cartwright, P., & Hughes, M. (2009). *Wind energy generation: Modelling and control.* West Sussex, England: John Wiley & Sons, Ltd.

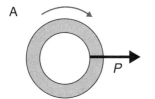
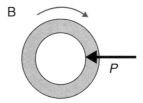

FIGURE 5-2 Schematic of DFIG supersynchronous (A) and subsynchronous (B) power delivery.
Adapted from Anaya-Lara, O., Jenkins, N., Ekanayake, J., Cartwright, P., & Hughes, M. (2009). *Wind energy generation: Modelling and control.* West Sussex, England: John Wiley & Sons, Ltd.

In that configuration, slip rings drive current into or out of the rotor winding. Thus, controllable voltage is driven to the rotor at slip frequency. Controllable voltage is key. It allows for variable-speed operations.

The rotor windings are looped through a variable-frequency power converter. These converters are usually based on two alternating current/direct current (AC/DC) voltage source converters. The converters are usually insulated gate bipolar transistors (IGBTs). The voltage source converters are linked via a DC bus.

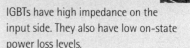

NOTE

IGBTs have high impedance on the input side. They also have low on-state power loss levels.

These connections enable variable-speed operations. The synchronous frequency of the grid is not linked directly to the rotor output. Rotor speeds carrying mechanical energy vary. Thus, these generators can function efficiently in varying wind conditions. Electric crowbars protect the electrical components of these generators. An **electric crowbar** is a circuit that intentionally shorts out. This happens if grid voltage surges above a certain level. They are named after the tool. If you dropped a real crowbar on electrical cables, the cables would probably disconnect. This is due to the weight of the heavy tool.

Doubly fed induction generators supply network power in two ways. The first is via the generator stator. The second is via the voltage source converters. The generator stator feeds power into the grid. This happens if the generator functions at above-synchronous speed. At times, the generator is functioning at below-synchronous speeds. Then, the voltage source converters feed the power into the grid. **FIGURE 5-2** shows a visual representation of these two states.

Doubly Fed Induction Generator Characteristics

Once a doubly fed induction generator is operative, it is in steady state. Its steady-state operation is based on the Steinmetz per-phase equivalent circuit model.

In **FIGURE 5-3**, stator and rotor voltages are represented by v_s and v_r. The stator and rotor currents are represented by i_s

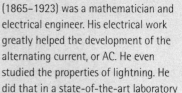

NOTE

Charles Proteus Steinmetz (1865–1923) was a mathematician and electrical engineer. His electrical work greatly helped the development of the alternating current, or AC. He even studied the properties of lightning. He did that in a state-of-the-art laboratory that was the size of an American football field.

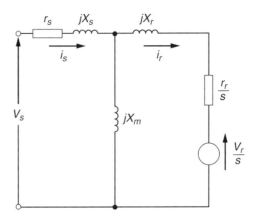

FIGURE 5-3 Steady-state operating characteristics of DFIGs.

Adapted from Anaya-Lara, O., Jenkins, N., Ekanayake, J., Cartwright, P., & Hughes, M. (2009). *Wind energy generation: Modelling and control.* West Sussex, England: John Wiley & Sons, Ltd.

and i_r. The per-phase stator and rotor resistances are represented by r_s and r_r. The stator and rotor leakage resistances are represented by X_s and X_r. Lastly, X_m represents magnetizing reactance and s represents slip. By transferring the magnetizing branch to the terminals, the equivalent circuit is simplified. See **FIGURE 5-4**.

Doubly fed induction generator torque-slip curves can be derived. They are derived from mathematical equations. Here is the formula for calculation of rotor current (I_r):

$$I_r = \frac{V_s - \left(\dfrac{V_r}{s}\right)}{\left(r_s + \dfrac{r_r}{s}\right) + j(X_s + X_r)}$$

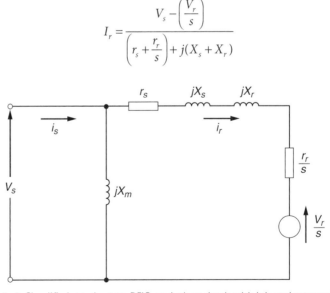

FIGURE 5-4 Simplified steady-state DFIG equivalent circuit with injected rotor voltage.

Adapted from Anaya-Lara, O., Jenkins, N., Ekanayake, J., Cartwright, P., & Hughes, M. (2009). *Wind energy generation: Modelling and control.* West Sussex, England: John Wiley & Sons, Ltd.

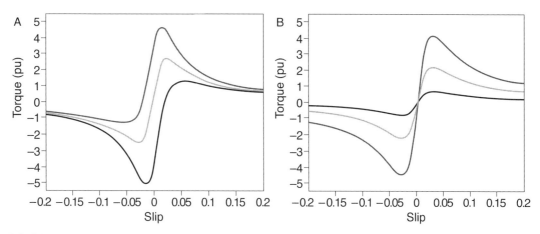

FIGURE 5-5 FSIG and DFIG torque-slip characteristics with (A) in-phase rotor injection and with (B) quadrature rotor injection.

Data from Anaya-Lara, O., Jenkins, N., Ekanayake, J., Cartwright, P., & Hughes, M. (2009). *Wind energy generation: Modelling and control.* West Sussex, England: John Wiley & Sons, Ltd.

A doubly fed induction generator's **electric torque** can be derived from the following formula. Electric torque (T_e) is the power balance in the gap between stator and rotor:

$$T_e = \left(I_r^2 \frac{\mathrm{r}_r}{\mathrm{s}} \right) + \frac{P_r}{\mathrm{s}}$$

P_r in the preceding equation represents **rotor active power**. Rotor active power is the amount of energy either supplied or removed by the voltage source converters. Here is the formula to derive rotor active power:

$$P_r = \frac{\mathrm{V_r}}{\mathrm{s}} I_r \cos\theta$$

This formula can also be calculated as such:

$$P_r = \mathrm{Re}\left(\frac{\mathrm{V_r}}{\mathrm{s}} \right) \mathbf{I}_r^*$$

FIGURE 5-5 shows the DFIG and fixed-speed induction generator (FSIG) torque-slip characteristics. It shows two forms of injection. The left side of the figure (*A*) shows in-phase rotor injection. The right side of the figure (*B*) shows quadrature rotor injection. V_{qr} represents the in-phase injection. V_{dr} represents the quadrature rotor injection characteristics.

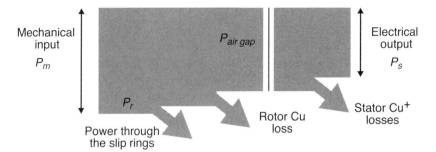

P_m = Mechanical power that wind delivers to a given turbine

P_r = Mechanical power from rotor to generator

$P_{air\ gap}$ = Power in the generator's air gap

P_s = Generator's stator power

FIGURE 5-6 Doubly fed induction generator power relationship diagram.

Adapted from Anaya-Lara, O., Jenkins, N., Ekanayake, J., Cartwright, P., & Hughes, M. (2009). *Wind energy generation: Modelling and control.* West Sussex, England: John Wiley & Sons, Ltd.

FIGURE 5-6 shows the relationship between rotor/stator electrical power and mechanical power in doubly fed induction generators.

Follow this series of equations closely. You will see how to calculate power fed into the grid by doubly fed induction generators:

1. Ignore stator power losses. You will calculate them later:

$$\textbf{Equation 1: } P_{air\text{-}gap} = P_s$$

2. Ignore rotor power losses. You will calculate them later:

$$\textbf{Equation 2: } P_{air\text{-}gap} = P_m - P_r$$

3. Ignoring both stator and rotor power losses, total power delivered to the grid is expressed by the following formula. This formula is a combination of Equations 1 and 2:

$$\textbf{Equation 3: } P_s = P_m - P_r$$

4. Express the preceding formula related to generator torque. Do so by considering the following conversions: $T\omega_s = P_s$ and $T\omega_r = P_m$.

$$\textbf{Equation 4: } T\omega_s = T\omega_r - P_r$$

5. Thus:

$$\textbf{Equation 5: } P_r = -T(\omega_s - \omega_r)$$

6. Separately, you calculate the stator and rotor power values. You do this by their relationships to slip. Here is the equation:

$$\textbf{Equation 6: } P_r = -sT\omega_s = -sP_s$$

7. Now you combine Equations 3 and 6 to calculate the mechanical power DFIGs produce:

Equation 7: $P_m = P_s + P_r = P_s - sP_s = (1 - s)P_s$

8. Finally, you calculate the true energy fed into the grid. Note that P_g represents this total in Equation 8:

Equation 8: $P_g = P_s + P_r$

Now you can see why the slip's controllable range is key. That value determines the size of necessary converters for doubly fed induction generators. The normal slip and speed range for most DFIGs is from 0.7 to 1.2 pu.

Doubly Fed Induction Generator Design

Doubly fed induction generators are similar to fixed-speed induction generators. **FIGURE 5-7** shows a cutaway diagram of a typical induction generator's components. Note that they are very similar to those of a fixed-speed induction generator. The main differences between these two generator types are detailed below.

The stator housings in a DFIG are steel and come in standard sizes. Turbine manufacturers set the design standards. The standards allow generators to fit inside the nacelle housing. It also has to fit with other turbine components.

Windings are copper cables laid into grooves in the stator and rotor (in wound-rotor designs). The cables are insulated. This provides protection from the environment. It also stabilizes their behavior.

The exterior of a DFIG, like that of a fixed-speed induction generator, is made to protect the insides. Its purpose is to protect inner parts from dew formation, dust particles, insects, and damage. A common design is the open drip-proof design. Another common design is the totally enclosed, fan-cooled (TEFC) design. The open drip-proof design is more common. This is because the nacelle housing doubles to protect inner parts from environmental dangers. However, studies show TEFC designs provide extra protection from damage. Thus, TEFC generator housings are becoming more popular in new designs.

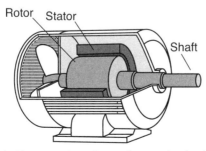

FIGURE 5-7 Typical induction generator cutaway diagram showing internal components.
Adapted from Manwell, J. F., McGowan, J. G., & Rogers, A. L. (2009). *Wind energy explained: Theory, design and application* (2nd ed.). West Sussex, UK: John Wiley & Sons, Ltd.

Doubly Fed Induction Generator Control

Doubly fed induction generator control depends on the power converter. It allows variable-speed operations. It does so by delinking rotor speed from power frequency. One system in use today uses the rotor-side power converter to control rotor torque. It works with the turbine's power factor or terminal voltage controls. The DC link voltage is under the control of the network-side power converter. The network-side power converter can also provide reactive power.

The control strategy is always based around maximum extraction of mechanical energy. However, it is impossible to extract wind power efficiently at all wind speeds. At low wind speeds, turbines operate at constant speed. This is affected by aerodynamic noise limits. In such cases, the controller allows torque to increase at constant speed. This continues until the rotor torque reaches its limited rating. Wind speeds can increase beyond this point. Rotor speed also increases, but *electromagnetic* torque is maintained in the generator. Pitch controls are used for torque control at very high wind speeds. If wind speeds reach dangerous levels, the turbine is shut down to prevent physical or personal damages.

Rotor Flux Magnitude and Angle Control

In a **rotor flux magnitude and angle control scheme**, generator terminal voltage and power output are modulated. This happens via adjustment of magnitude and angle of the **rotor flux vector**. Rotor flux vector is a characteristic of electromagnetization. Generators that contain permanent magnets have fixed rotor flux vectors, but in electromagnetized generators, the rotor flux vector can be changed. It is based on controlling stator currents and rotor slip. This control method has advantages. One is minimal interaction between the voltage and power control circuits. Another is increased system damping and voltage recovery functions. Voltage recovery functions are vital after system faults.

In **FIGURE 5-8**, note the two distinct loops. One loop controls terminal voltage. The other controls generator power output. Either the internal voltage vector or the rotor flux vector can be the control vector. This is because they are directly related to each other. In the following paragraphs, consider the internal voltage vector to be the control vector.

The voltage control circuit forms an error signal. The error signal results from the difference in terminal voltage and its specified reference value. The error signal is sent to the automatic voltage regulator (AVR) compensator. The compensator processes the signal. Then it adjusts the internal voltage vector magnitude. It aligns that value with its desired reference value.

The power control circuit works based on different inputs. The inputs come from the power-speed characteristic. The power-speed characteristic is based on wind speeds. The difference between the generator reference set point and its

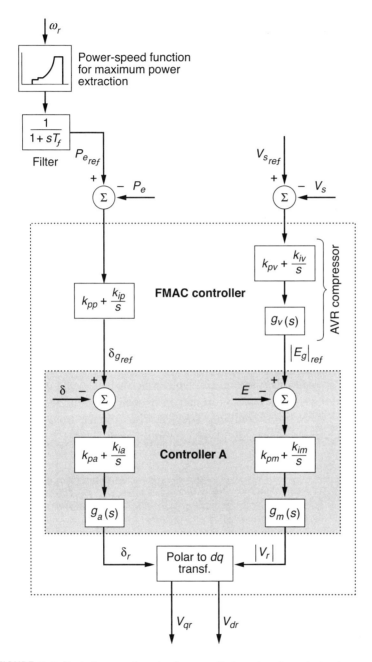

FIGURE 5-8 Block diagram of a rotor flux magnitude and angle control system.

Adapted from Anaya-Lara, O., Jenkins, N., Ekanayake, J., Cartwright, P., & Hughes, M. (2009). *Wind energy generation: Modelling and control.* West Sussex, England: John Wiley & Sons, Ltd.

actual power level forms an error signal. The AVR compensator processes the error signal. The compensator initiates the correct reference value for the control vector's angular position. That reference value is based on stator voltage vector.

PI controllers are present in both voltage and power control circuits. The voltage control circuit also uses lead-lag compensation. This ensures correct response speed. It also stabilizes individual loop margins.

Lastly, the rotor voltage vector transforms. It transforms from polar to rectangular coordinates. The rectangular coordinates are sent to the pulse-width modulation (PWM) generator. The PWM generator uses this information. The information helps the PWM generator control the rotor-side converter's switching.

Current-Mode Torque Control

Current-mode torque controllers change electromagnetic torque in doubly fed induction generators. In this technique, the rotor current is divided into two orthogonal components. The components are called d and q. The q component regulates torque, and the d component controls terminal voltage. They do this in reaction to wind speed changes, to optimize generator operations. Thus, the controllers direct the system to its necessary operating point reference. At any given rotor speed, controllers change wind turbine characteristics for maximum power extraction. Proportional-integral (PI) controllers set the rotor voltage needed for maximum extraction. They do so together with a compensation term. These functions combine. They can then reduce cross-coupling between speed and voltage control loops.

NOTE

In control systems, a proportional-integral controller is called a PI controller. It controls voltage via weighted error sums. Weighted error sums come from the difference between actual output and desired set point. The integral of that value drives the controller to adjust voltage.

MECHANICS

Even though this course focuses on many electronic principles, it is important to understand mechanics, too. According to *The New Oxford American Dictionary*, mechanics is "the branch of applied mathematics dealing with motion and forces producing motion." In other words, mechanics is a branch of mathematics. It is related to the causes of all motion. This is applicable to wind energy. The initial extraction of power from wind is via mechanics. After the mechanical transfer of power to the generator, the energy is converted to electricity. Then it is fed into the grid via electrical methods.

There are many categories of mechanics. The following is a list of some common branches, including links to some websites where you can find more information on different fields of

MECHANICS (Continued)

mechanics. (Most of these links are to a reputable website with references, Science Daily. The Wikipedia link is for familiarity purposes. Most Wikipedia entries can be edited by anyone. Therefore, you should consult with your instructors about possible errors in Wikipedia articles.)

- General mechanical principles—*http://en.wikipedia.org/wiki/Mechanics*
- Fluid mechanics—*www.sciencedaily.com/articles/f/fluid_mechanics.htm*
  Fluid mechanics is the study of the movement of fluids. The term *fluids* includes all liquids and gases. From those categories, several more are derived.
- Shear stress—*www.sciencedaily.com/articles/s/shear_stress.htm*
  Shear stress is a state where materials change shapes. The shape changes usually result from torque. Torque is a mechanical concept you should understand. The shape changes do not imply any change in volume.
- Momentum—*www.sciencedaily.com/articles/m/momentum.htm*
  Momentum is a physical concept related to the mass and volume of an object. For example, imagine two identical-mass trucks moving at different speeds. The slower truck has less momentum than the faster truck. Now imagine a small car traveling next to a large truck at the same speed. The large truck has more momentum than the small car.
- Quantum tunneling—*www.sciencedaily.com/articles/q/quantum_tunnelling.htm*
  Quantum tunneling is a specialized area of mechanics. It focuses on the behavior of particles at subatomic levels. Consider a car climbing a hill. If the car does not have enough speed, it will not make it over the hill. However, at subatomic levels, the same "car" can make it over the hill. Or it could disappear from the hill and appear on the other side. Do these results seem impossible? If so, read about quantum tunneling. Scientists are researching today how this type of physical behavior is possible.

CHAPTER SUMMARY

In this chapter, you continued to expand your knowledge of generators. Specifically, you focused on doubly fed induction generators. They are similar to fixed-speed induction generators, but there are some important differences. You started by studying basic workings of doubly fed induction generators. This includes mechanisms that permit variable-speed operations. You also covered two methods these generators use to feed power into a grid. Next, you covered the characteristics of doubly fed induction generators. These revolve around mathematical principles. Using these principles, you reviewed the method for calculating doubly fed induction generator power generation. Then you covered doubly fed induction generator design. Doubly fed induction generator design is close to that of fixed-speed induction generators. The last section explained two methods of power and voltage control. The first model is the most common. It is the rotor flux magnitude and angle control method. The second method is the current-mode torque control method.

CHAPTER KEY CONCEPTS AND TERMS

Electric crowbar
Electric torque
Rotor active power
Rotor flux magnitude and angle control scheme
Rotor flux vector

CHAPTER ASSESSMENT: DOUBLY FED INDUCTION GENERATORS

1. DFIGs provide the most efficient way of converting wind energy to electricity.
   ❑ **A.** True
   ❑ **B.** False

2. Doubly fed induction generators use a(n) _____ configuration.

3. Doubly fed induction generators' rotor configuration allows for _____ operation.

4. Doubly fed induction generators supply network power in two ways. The first is via the _____. The second is via the _____.

5. The amount of energy either supplied or removed by the voltage source converters is called:
   ❑ **A.** electric torque.
   ❑ **B.** rotor active power.
   ❑ **C.** stator active power.
   ❑ **D.** electromagnetic torque.

6. What does the formula $T_e = \left(P_r \dfrac{r_r}{s} \right) + \dfrac{P_r}{s}$ derive?
   - ❏ **A.** Electric torque
   - ❏ **B.** Rotor active power
   - ❏ **C.** Stator active power
   - ❏ **D.** Electromagnetic torque

7. Which formula is used to calculate the total power fed into the grid?
   - ❏ **A.** $P_r = -sT\omega_s = -sP_s$
   - ❏ **B.** $P_g = P_s + P_r$
   - ❏ **C.** $T\omega_s = T\omega_r - P_r$
   - ❏ **D.** $P_{air\text{-}gap} = P_m - P_r$

8. The normal slip and speed range for most DFIGs is from 0.5 to 1.2 pu.
   - ❏ **A.** True
   - ❏ **B.** False

9. Windings are made of _____ cables laid into grooves in the stator and rotor.

10. _____ _____ allow variable-speed operations by delinking rotor speed from power frequency.

11. DFIGs extract wind power efficiently at all wind speeds.
    - ❏ **A.** True
    - ❏ **B.** False

12. Current-mode torque controllers change _____ _____ in doubly fed induction generators.

13. Which of the following are modulated in a rotor flux magnitude and angle control scheme?
    - ❏ **A.** Terminal voltage and power output
    - ❏ **B.** Terminal voltage and electromagnetic torque
    - ❏ **C.** Rotor-side power conversion and power output
    - ❏ **D.** Slip and torque

14. PI controllers are present only in voltage control circuits.
    - ❏ **A.** True
    - ❏ **B.** False

15. The PWM generator controls the rotor-side converter's switching based on which of the following?
    - ❏ **A.** Polar coordinates
    - ❏ **B.** Rectangular coordinates
    - ❏ **C.** Polar and rectangular coordinates
    - ❏ **D.** None of the above

Fully Rated Converter-Based Generators

IN THIS CHAPTER, you will continue building your knowledge of generators. Fully rated converter-based generators (FRC generators) are the most flexible of all generator types. They can be conventional, synchronous, or asynchronous. They also utilize electromagnetic *or* permanent magnet-based excitation. Each type of excitation has benefits and drawbacks.

The chapter opens with an overview of FRC generators. You will read about what distinguishes these from other types of generators. Next, you will study design and characteristics of FRC generators. For the most part, FRC generators are designed like most other generators, but there are some key differences. You will focus on direct-drive FRC generators. These generators contain no gearbox between the rotor and the generator. You will also study some differences in permanent magnet-based versus electromagnetic excitation. The last section of this chapter exposes you to two different control methods. FRC generators can be controlled with two different control methodologies. The first is the load angle method. The second method is called the vector control strategy. You will study both methods in this chapter. FRC generators are becoming more popular in wind energy applications. That is because they are flexible machines. They can be adapted for many different circumstances. Read on to learn more about FRC generators.

Chapter Topics

This chapter covers the following topics and concepts:

- FRC generator overview
- FRC generator characteristics and design
- FRC generator control

Chapter Goals

When you complete this chapter, you will be able to:

- Compare and contrast FRC synchronous and FRC induction generator characteristics
- Discuss the main advantages and drawbacks of direct-drive generators
- Diagram similarities and differences between electromagnetic and permanent magnet-based excitation
- Explain why early generator manufacturers transitioned from permanent magnet-based to electromagnetic excitation
- Calculate active power and reactive power values given the necessary values
- Discuss the basics of load angle control for FRC generators
- List the benefits of the vector control methodology for FRC generators

FRC Generator Overview

FRC generators are so named because they contain fully rated voltage source converters (VSCs). This permits them to operate at full power. Converters control them. They feature variable-speed operations. Because of these benefits, fully rated VSCs fulfill current grid requirements. Grid requirements revolve around issues such as the following:

- Fault ride-through requirements
- Active power and frequency control
- Voltage and frequency operating ranges
- Reactive power control and voltage regulation

Active power is the amount of power generated for load consumption. **Reactive power** is the resultant power loss derived from power generation. Reactive power is used in different ways.

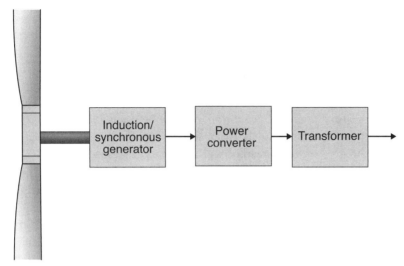

FIGURE 6-1 Schematic diagram of an FRC generator.

Adapted from Anaya-Lara, O., Jenkins, N., Ekanayake, J., Cartwright, P., & Hughes, M. (2009). *Wind energy generation: Modelling and control*. West Sussex, England: John Wiley & Sons, Ltd.

See **FIGURE 6-1**. It shows that FRC generators are similar in design and functionality to induction generators, but their construction allows for extra flexibility. Technological advances permit operators to make "custom-designed" generators for their needs. FRC generators can be of the conventional, asynchronous, or synchronous type. Permanent magnets can replace electromagnetic windings. Some FRC generators do not require a gearbox. All rotor power is transferred through a power converter. So the operational characteristics and dynamics are isolated from the power grid. This isolation permits generator frequency variation. This is necessary for variable-speed operations. Variable-speed operations are useful in changing wind speeds. The generator's rated power determines the power converter rating.

There are multiple configuration options for power converters. A generator-side converter could consist of a diode-based rectifier. Or it could be a pulse-width modulation, variable-speed control (PWM-VSC) component. On the other hand, network-side converters are usually PWM-VSCs. The power-converter arrangement determines which operation and power-control strategy is best for a turbine.

FRC Generator Characteristics and Design

Synchronous FRC generators can be magnetized in two ways. The most common way is electromagnetic. However, another configuration contains permanent magnets. In a direct-drive configuration, the generator receives mechanical energy from the rotor. This comes via the main shaft from the rotor. It does not use a gearbox. These generators are designed for low-speed operations. They also tend to be large. This is because they contain a high number of poles. Many FRC turbines use a single-stage, low-ratio gearbox though. This allows use of a smaller generator with fewer poles.

Direct-Drive Generators

Currently, most turbines rated more than just a few kilowatts use four-pole (standard) generators. Standard generators operate between 750 and 1,800 rpm. Turbine speed is usually much lower. Rotors usually turn at 20 to 60 rpm. Therefore, in most turbines, gearboxes provide the torque/speed conversion necessary for high-rpm generators. Alternatively, a direct-drive generator can be used. See the schematic diagrams in **FIGURE 6-2**. They show layouts of conventional and direct-drive generator wind turbines.

Direct-drive generators have two main benefits. The first is a reduction in drive train power losses. This is because no gearbox is used. Gearboxes tend to reduce power transfer from rotor to the generator. Because there is no gearbox to maintain, costs are lower and revenue is higher. The second benefit is that direct-drive turbines are quieter than conventional ones. This is largely because there is no gearbox.

Using no gearbox increases torque. So direct-drive generators must have very high torque ratings. Generator size and power losses depend on torque rating, not power rating. For example, take a 500 kW, 30 rpm generator and a 50 MW, 3,000 rpm generator. Both generators should have the same torque rating.

Due to high torque rating and a large number of poles, direct-drive generators are larger than those that use a gearbox. They are also less efficient. Thus,

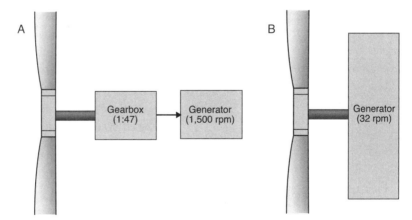

FIGURE 6-2 Schematic diagram of a conventional turbine generator configuration (A) versus a direct-drive configuration (B).

Adapted from Anaya-Lara, O., Jenkins, N., Ekanayake, J., Cartwright, P., & Hughes, M. (2009). *Wind energy generation: Modelling and control.* West Sussex, England: John Wiley & Sons, Ltd.

direct-drive generator design utilizes a large diameter. That reduces the weight of active parts. Direct-drive design also depends on small **pole pitch**. Pole pitch is the distance between poles based on angular separation. Reduced pole pitch helps to maximize generator efficiency.

Permanent Magnet Versus Electromagnetic Excitation

Early generators used permanent magnets for **excitation**. Excitation is the interaction between static and rotating magnetic poles. Technology advances permitted electromagnetic excitation. Early permanent magnets were heavy. So most manufacturers started using electromagnetic excitation.

Synchronous generators can self-excite. They use copper windings. Copper windings generate electromagnetic fields. Synchronous generators can also use permanent magnets. They create a "real" magnetic field. Wound-rotor generators have an advantage. The excitation current is adjustable. This permits delinking output voltage control from load current levels. So most constant-speed generators use wound rotors. This is especially true in direct-grid connections, but most synchronous generators connect to a grid with electronic converters. Then delinking output voltage control from load current is less beneficial.

Nowadays wound-rotor generators are heavier than permanent magnet-based versions. Generators with small pole pitch are also bulkier. Copper windings contribute to power losses. That is because each pole needs windings. As the number of poles increases, the number of windings increases. As winding quantities increase, power losses grow. Permanent magnets do lose some power. Permanent magnet losses are lower than copper winding losses.

> **NOTE**
>
> A dual-pole machine has two poles: one N and one S. They are arranged on a circular axis and are therefore equidistant from each other. In this simple configuration, pole pitch is 180 degrees. So in a standard generator, pole pitch is 90 degrees.

> **NOTE**
>
> Magnets are made from materials that produce magnetic fields. Traditionally this was iron. Iron is a heavy metal, but many materials are attracted to magnets. Any of them can become permanent magnets. These materials are called *ferromagnetic*. Currently, magnets are made from iron *and* lighter metals. They are made of nickel, cobalt, and other alloys. They can even be made of minerals. Lodestone is one example.

Permanent Magnet Synchronous Generators

Permanent magnet-based generators have advantages. One advantage is that no field current supply is needed. Another is that permanent magnets require no reactive power compensation. Permanent magnets do not require slip rings.

FIGURE 6-3 . In this generator configuration, a DC booster stabilizes DC link voltage. A grid-side converter (PWM-VSC) controls generator operations. There are two methods for PWM-VSC control. The first is with load angle methods. The second is with voltage-oriented reference frame current controllers. The maximum-power/speed characteristic defines the power reference. **FIGURE 6-4** . It represents the maximum-power/speed characteristic.

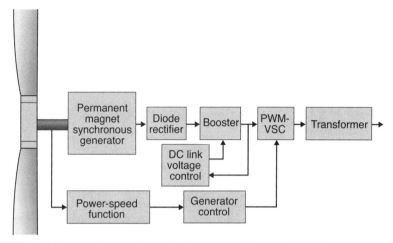

FIGURE 6-3 Permanent magnet-based FRC generator with uncontrolled diode-driven, generator-side converter.

Adapted from Anaya-Lara, O., Jenkins, N., Ekanayake, J., Cartwright, P., & Hughes, M. (2009). *Wind energy generation: Modelling and control.* West Sussex, England: John Wiley & Sons, Ltd.

Some generators with permanent magnets use two VSCs. They are on the grid side. They have a back-to-back layout. In this layout, each PWM-VSC does a different job. The generator-side PWM-VSC controls generator functionality. The network-side generator controls DC link voltage. It feeds excess active power into the grid. **FIGURE 6-5** is a diagram of this configuration.

FRC Induction Generators

Refer back to Figure 6-1. FRC induction generators (FRC-IGs) operate at variable frequencies. See **FIGURE 6-6** for FRC-IG steady-state characteristics. The characteristics are based on three measurement sets. The first is torque/rotor speed. The second is active power/reactive power. The third is slip/reactive power.

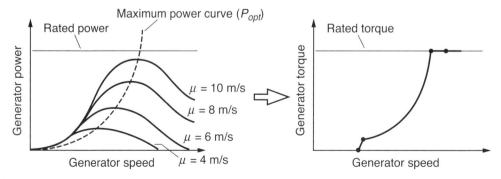

FIGURE 6-4 FRC generator characteristic for maximum wind energy extraction.

Data from Anaya-Lara, O., Jenkins, N., Ekanayake, J., Cartwright, P., & Hughes, M. (2009). *Wind energy generation: Modelling and control.* West Sussex, England: John Wiley & Sons, Ltd.

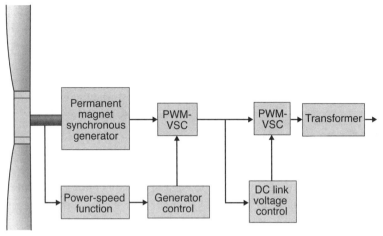

FIGURE 6-5 Permanent magnet-based FRC generator using two back-to-back PWM-VSC voltage source converters.

Adapted from Anaya-Lara, O., Jenkins, N., Ekanayake, J., Cartwright, P., & Hughes, M. (2009). *Wind energy generation: Modelling and control.* West Sussex, England: John Wiley & Sons, Ltd.

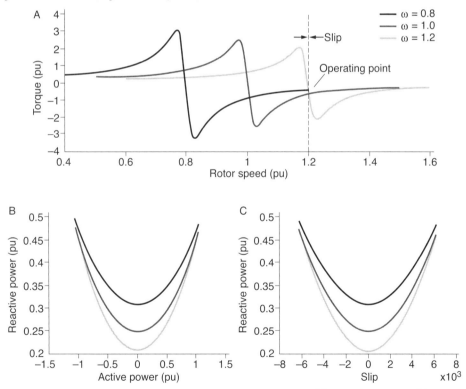

FIGURE 6-6 FRC-IG steady-state operating characteristics as measured by (A) torque/rotor speed, (B) active power/ reactive power, and (C) slip/reactive power.

Data from Anaya-Lara, O., Jenkins, N., Ekanayake, J., Cartwright, P., & Hughes, M. (2009). *Wind energy generation: Modelling and control.* West Sussex, England: John Wiley & Sons, Ltd.

Now refer back to Figure 6-4. Generator speed must vary along with wind speed. This allows for the maximum power extraction presented in that figure. The PWM network-side converter control signal changes to allow this outcome. At low wind speeds, generator frequency is low. At high wind speeds, the generator frequency increases. Reactive power absorbed by the generator is constant at all frequencies. Slip permits stabilization of reactive power absorption. Maximum operating speed is limited to 1.2 pu. So generator operating frequency is also capped at 1.2 pu.

NOTE

As FRC technology improves, the ability to better control power generated is increasing and the total power output is increasing. FRC generator control may be the most important development in wind power–based generation, but FRC generator control is not new in wind turbines, as it has been used for several decades in wind turbines and was used previously in hydroelectric generation.

FRC Generator Control

There are two control methods for synchronous FRC generators. The first is load angle control. It controls generator-side converters. The second method is vector control. It controls network- and generator-side converters.

Load Angle Control

Load angle control depends on steady-state electricity flow **FIGURE 6-7**. Equations 1–4 can calculate active and reactive power transfer. These transfers are between the DC link and the generator.

These load angle equations are for steady-state operations:

Equation 1 (active power flow): $P = \dfrac{E_g B_t}{X_g} \sin\alpha_g$

Equation 2 (reactive power flow): $H = \dfrac{E_g{}^2 - E_g B_t \cos\alpha_g}{X_g}$

E_g = Generator internal voltage magnitude
X_g = Synchronous reactance
V_t = Converter terminal voltage magnitude
α_g = Phase differential between E_g and V_t

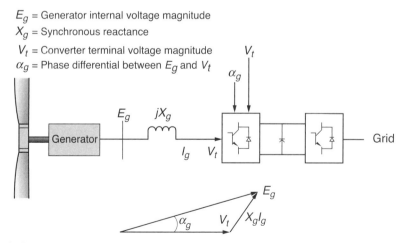

FIGURE 6-7 Schematic diagram of synchronous generator control via load angle method.
Adapted from Anaya-Lara, O., Jenkins, N., Ekanayake, J., Cartwright, P., & Hughes, M. (2009). *Wind energy generation: Modelling and control.* West Sussex, England: John Wiley & Sons, Ltd.

Load angle is usually negligible. Thus these formulas are simplified in Equations 3 and 4:

Equation 3 (active power flow [simplified]): $P = \dfrac{E_g B_t}{X_g} \alpha_g$

Equation 4 (reactive power flow [simplified]): $H = \dfrac{E_g{}^2 - E_g B_t}{X_g}$

These equations show two concepts. The first is that phase angle determines active power transfer rates. The second is that voltage magnitudes control reactive power transfer rates. Reactive power flows from higher-magnitude points to lower-magnitude points.

Adjusting magnitude and angle of the generator-side converter AC terminals controls two items. The first is generator functionality. The second is power transfer from the generator to the DC link. You calculate the magnitude (B_t). You can also find the angle. Use Equations 5 and 6, where $P_{g_{ref}}$ = active power transfer needs from generator to DC link and $H_{g_{ref}}$ = reactive power reference value.

Equation 5 (magnitude): $B_t = E_g - \dfrac{H_{g_{ref}} X_g}{E_g}$

Equation 6 (angle): $\alpha_g = \dfrac{P_{g_{ref}} X_g}{E_g B_t}$

The main advantage of load angle control is lack of complication, but it does not account for generator dynamics. Thus, load angle control might not be effective in variable conditions. See **FIGURE 6-8** for a generic diagram of a generator-side converter load angle control scheme.

Vector Control

A dynamic framework in synchronous generators determines vector control methods. It is called the *dq* framework. The framework is the *d* axis. It is aligned with the rotor's magnetic axis. Remember that vector control strategies control for network-side and generator-side converters.

A controller determines two reference currents in this methodology. Once you know these, you can calculate the voltage magnitudes. You calculate voltage magnitudes with these equations:

Equation 7: $\bar{v}_{ds} = -\bar{\rho}_s \bar{\iota}_{qs} + \bar{X}_{qs} \bar{\iota}_{qs}$

Equation 8: $\bar{v}_{qs} = -\bar{\rho}_s \bar{\iota}_{qs} - \bar{X}_{ds} \bar{\iota}_{ds} + \bar{E}_{fd}$

As shown in **FIGURE 6-9**, torque control is on the *q* axis. Generator magnetization is on the *d* axis. Normally there is an error between the reference and

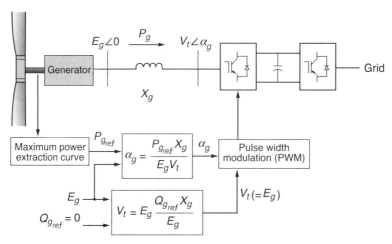

FIGURE 6-8 Implementation of generator-side converter load angle control.

Adapted from Anaya-Lara, O., Jenkins, N., Ekanayake, J., Cartwright, P., & Hughes, M. (2009). *Wind energy generation: Modelling and control.* West Sussex, England: John Wiley & Sons, Ltd.

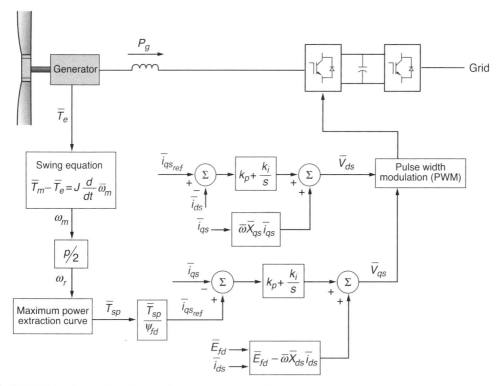

FIGURE 6-9 Implementation of generator-side converter vector control.

Adapted from Anaya-Lara, O., Jenkins, N., Ekanayake, J., Cartwright, P., & Hughes, M. (2009). *Wind energy generation: Modelling and control.* West Sussex, England: John Wiley & Sons, Ltd.

actual current values. To correct for this, vector control methods utilize PI controllers. The PI controller sends voltage to the q axis. The voltage is the necessary amount for generator-side converter control. There is also an error between the reference and actual stator currents. These currents are in the d axis. A PI controller processes this error, too. Then it corrects for the error in actual stator current in the d axis. The reference value for stator current is 0 in synchronous generators with permanent magnets.

TURBULENCE

If you've ever flown in any kind of aircraft, you have probably experienced the effects of turbulence—the "bumps" you can feel sometimes when regular airflow over the wing is interrupted. Turbulence is a characteristic of flow in fluid dynamics. It is defined by chaotic, random property changes in flowing fluids. Because air is defined as a fluid for scientific purposes, air is subject to turbulent behavior. Both airplane wings and turbine blades must be designed to withstand certain turbulence limits. Some different kinds of random fluid property changes include the following:

- Low-momentum diffusion
- High-momentum convection
- Rapid pressure fluctuations
- Unstable wind velocities

Consider how water flows over a simple, smooth object such as a round stone. If the water is flowing slowly, its flow is smooth with few to no irregularities. This situation is called laminar flow. If you increase the flow, the water will begin to show irregularities as it flows. If you increase the flow even more, the irregularities will begin to dominate the flow pattern. The irregularities that you see are the turbulence in the water flow.

Other related factors can influence the laminar to turbulent fluid characteristic. Large objects, for example, cause more turbulence than small objects. Decreasing fluid viscosity and increasing fluid density are two other causes of increased turbulence.

Turbulence is a major issue in wind energy. This is especially true as wind turbine sizes have grown over the decades. You will focus more attention on turbulence and its effects in future wind energy training courses.

Recently, researchers have started to look at using the turbulence, particularly of vertical axis wind turbines (VAWTs), as constructive interference waves to enhance the output of wind farms. The principle is similar to that of a school of fish that swim together using less energy than they would if not in their well-formed school.

CHAPTER SUMMARY

In this chapter, you added to your knowledge of generators. FRC generators are the most flexible generator types. They have conventional, synchronous, or asynchronous functionalities. They also use electromagnetic *or* permanent magnet-based excitation. Each type of excitation has benefits and drawbacks. FRC generators can have high torque ratings. They don't necessarily require a gearbox.

The chapter started with an overview of FRC generators. You studied what distinguishes these generators from others. Then you studied design and characteristics of FRC generators. FRC generators are designed like most other generators, but there are some key differences. For example, you focused on direct-drive FRC generators. These generators require no gearbox. Thus their torque ratings are sometimes very high. You also covered the differences in permanent magnet-based versus electromagnetic excitation. The last section of this chapter focused on control. It explained two different FRC generator control methods. The first is the load angle method. The second is the vector control strategy. Because they are flexible machines, adaptable for different circumstances, FRC generators are becoming more and more popular in wind energy.

CHAPTER KEY CONCEPTS AND TERMS

Active power
Excitation
Pole pitch
Reactive power

CHAPTER ASSESSMENT: FULLY RATED CONVERTER- BASED GENERATORS

1. Fully rated converter-based generators have which of the following?
    - ❑ **A.** Low-torque ratings
    - ❑ **B.** Fully rated VSCs
    - ❑ **C.** High torque ratings
    - ❑ **D.** Fully rated vector control

2. Which of the following definitions describes active power?
    - ❑ **A.** The amount of power needed for generator operations.
    - ❑ **B.** The amount of power generated for control vector strategies.
    - ❑ **C.** The amount of power generated during network voltage dips.
    - ❑ **D.** The amount of power generated for load consumption.

3. Reactive power is the resultant _____ _____ derived from power generation.

4. FRC generator converters can consist of which of the following?
   - ❏ **A.** PWM-VSCs
   - ❏ **B.** Diode-based rectifiers
   - ❏ **C.** A and B
   - ❏ **D.** None of the above

5. How many magnetization options exist for FRC generators?
   - ❏ **A.** One
   - ❏ **B.** Two
   - ❏ **C.** Three
   - ❏ **D.** Four

6. Direct-drive generators increase the power transfer from rotor to generator.
   - ❏ **A.** True
   - ❏ **B.** False

7. Direct-drive generators require _____ torque ratings than gearbox-driven generators.

8. Reduced pole pitch leads to which of the following?
   - ❏ **A.** Better generator efficiency
   - ❏ **B.** Heavier generator weight
   - ❏ **C.** Greater winding power losses
   - ❏ **D.** All of the above

9. Permanent magnet-based generators rely on field current supply to generate their electromagnetic fields.
   - ❏ **A.** True
   - ❏ **B.** False

10. In a back-to-back PWM-VSC configuration, the network-side PWM-VSC controls _____ _____ _____.

11. Maximum FRC-IG operating frequency is capped at which of the following?
    - ❏ **A.** 0.9 pu
    - ❏ **B.** 1.3 pu
    - ❏ **C.** 1.7 pu
    - ❏ **D.** None of the above

12. What does the following formula represent?

$$P = \frac{E_g B_t}{X_g} \sin \alpha_g$$

   - ❏ **A.** Nonsimplified equation for active power flow
   - ❏ **B.** Simplified equation for active power flow
   - ❏ **C.** Nonsimplified equation for reactive power flow
   - ❏ **D.** Simplified equation for reactive power flow

13. Reactive power flows from higher-magnitude points to lower-magnitude points.
   ❑ **A.** True
   ❑ **B.** False

14. A dynamic framework in synchronous generators determines _____ control methods.

15. In the vector control methodology, generator magnetization is centered on the _____ axis.

Wind Turbine Control

IN THIS CHAPTER, you will focus on wind turbine control mechanisms. Controllers allow wind turbines to operate efficiently. Sometimes there are control strategies that do the same. Control strategies consist of hardware that depends on manipulation to optimize wind power extraction.

This chapter opens with a discussion of frequency control. Frequency control is important because an entire grid must maintain a certain frequency to service all loads. The frequency cannot be too high or low at any given time. Next, you will focus on active stall wind turbines. These are usually fixed-speed operations. Thus, control methods allow for changes in frequency. It is important for active stall wind turbine operation flexibility. Then, you will study variable pitch angle control methods. These methods depend on pitchable blades. Blades change pitch to respond to frequency concerns. This is another way to allow for efficient turbine operations. Last, you will read about full rated power electronic converters and then topological wind system considerations. Wind system topology is also important because geographical features affect how power is generated. They also affect how it is transmitted and distributed. You will compare four common topological wind system layouts.

Chapter Topics

This chapter covers the following topics and concepts:

- Important aspects of frequency control
- Active stall wind turbine control methods
- Variable pitch angle control methods
- Full rated power electronic control strategies
- Wind system topological considerations

Chapter Goals

When you complete this chapter, you will be able to:

- List several key concerns regarding electrical system frequency
- Describe the allowable variations in common electrical system frequency
- Explain the concepts of occasional service, primary response, and secondary response
- Discuss the workings of active stall control systems
- Describe the basic mechanisms of pitch angle control systems
- Compare and contrast DFIG and FSIG implementations of pitch angle control methods
- List three reasons why topology in wind systems is a key consideration in wind farm design
- Chart four different types of wind system layouts, comparing metrics between each different wind system design

Frequency Control

In alternating current (AC) electrical systems, generation and consumption must be balanced constantly. Any imbalance can cause frequency disruptions. When load increases, frequency decreases. When load decreases, frequency increases. Acceptable limits for frequency changes are within a margin of +/– 1 percent. For example, the normal English and Welsh system frequency is 50 Hz. The allowable deviation in those countries is +/– 0.2 Hz. **Droop governors** often work to control generator frequency. They do this by decreasing generator equilibrium speed when load increases, but during unexpected large load or generation surges, the operable frequency margins increase. In the English and Welsh system, the absolute limits are +0.5 Hz and –0.8 Hz.

NOTE

North America has a 60 Hz frequency, and margins vary from country to country.

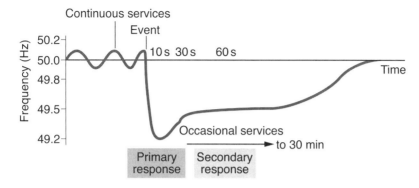

FIGURE 7-1 English and Welsh electrical system frequency control.

Data from Anaya-Lara, O., Jenkins, N., Ekanayake, J., Cartwright, P., & Hughes, M. (2009). *Wind energy generation: Modelling and control.* West Sussex, England: John Wiley & Sons, Ltd.

Sometimes generation drops, or large loads connect to the system. When this happens, frequency starts to drop. It drops at a rate equal to the entire system's angular momentum. The system's angular momentum is the totality of all generators' angular momentum plus the angular momentum of all spinning loads connected to the grid. In England and Wales, when frequency drops more than 0.2 Hz, additional generation capacity is contracted. This additional generation capacity is called **occasional service**. Occasional service usually consists of response time and duration. It is power delivered no more than 10 seconds after a frequency drop. It is expected to last from 20 seconds to 30 minutes after initial response time. Occasional service is divided into two responsibility categories. The first is called primary response. The second is called secondary response. Primary response comes from an automatic droop control loop. Generators produce more electricity when voltages fall outside of their **dead band**. Dead band is defined as a voltage range in the governor where no power output changes are possible. Their increased output also depends on their individual rotor-stator response times. Secondary response consists of normal frequency restoration with a supplementary control loop. These responses are represented in **FIGURE 7-1**.

Active Stall Wind Turbines

Active stall wind turbines operate at fixed speed. They use fixed-speed induction generators (FSIGs). FSIG wind turbines respond like synchronous machines when frequency drops. When frequency drops, machine speed decreases. The result is a surge of kinetic energy. The energy is converted to electricity. This electricity causes a temporary power surge. When frequency rises, the opposite generator reaction occurs. You can calculate the rotating machine mass. It is based on the next few equations. In these equations, ω = speed, E_k = kinetic energy, I = moment of inertia, and τ = torque.

$$\textbf{Equation 1: } E_k = \frac{1}{2}I\omega^2$$

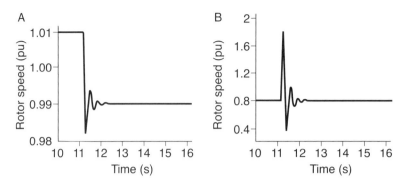

FIGURE 7-2 FSIG output change based on rotor speed (pu)/time (A) and output (pu)/time (B).

Data from Anaya-Lara, O., Jenkins, N., Ekanayake, J., Cartwright, P., & Hughes, M. (2009). *Wind energy generation: Modelling and control.* West Sussex, England: John Wiley & Sons, Ltd.

When ω changes, you can calculate extracted power (P) with Equation 2:

$$\textbf{Equation 2: } P = \frac{dE_k}{d\tau} = \frac{1}{2} I \times 2\omega \frac{d\omega}{d\tau} = I\omega \frac{d\omega}{d\tau}$$

FIGURE 7-2 shows output and speed changes in fixed-speed induction generators. It is based on a frequency step of 1.0 Hz. Fixed-speed wind turbines with ratings higher than 1 MW are inertially constrained. The constraints range from 3 to 5 seconds. This displays how quickly some wind turbines can respond to frequency changes.

> **NOTE**
>
> Many wind turbines in fact combine many methods for controlling their rpm and shutting down the turbine altogether. It is important to have redundant systems in case of failure, to prevent runaway conditions.

Doubly fed induction generators (DFIGs) with conventional controls restrain rotor torque. The restraint depends on an indexed curve based on rotor speed. There is no link between rotor speed and system frequency. Thus no inertial constraint is necessary. See **FIGURE 7-3**. It represents DFIG speed and output change in a frequency step of 1.0 Hz.

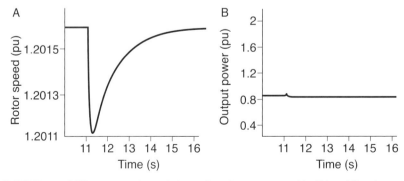

FIGURE 7-3 DFIG output and speed change based on rotor speed (pu)/time (A) and output (pu)/time (B).

Data from Anaya-Lara, O., Jenkins, N., Ekanayake, J., Cartwright, P., & Hughes, M. (2009). *Wind energy generation: Modelling and control.* West Sussex, England: John Wiley & Sons, Ltd.

Variable Pitch Angle Control

FSIG and DFIG turbines are unloadable via the **pitch angle power control** method. Pitch angle power control is useful with fixed-speed turbines. It depends on standard power control for feathering turbine blades. In this control method, the blade pitch is gradually reduced as wind speed increases. This helps maintain rated power output. When the turbine is operating above rated power, pitch control methods allow for wind energy **spilling**. Spilling is a name for the intentional reduction of wind power extraction. Spilling leaves a margin for extra wind turbine loading. That helps in providing **slow primary response**. Slow primary response is a certain form of occasional services. In slow primary response, there is no need for immediate turbine response to load changes. This is because the slight excess power available is driven to more transient loads. So no control system action is needed in slow primary response.

NOTE

Variable pitch angle control is the most common method for controlling wind turbine performance. However, this method requires more mechanical maintenance and is more prone to failure due to the increased complexity and stress on the rotor hub.

Below rated power, blade pitch is usually fixed for optimal energy extraction. The fixed angle is usually about –2 degrees. This angle varies in variable-speed turbines. However, it does not normally vary in fixed-speed turbines.

FIGURE 7-4 shows some wind power extraction curves. They are comparisons between different angles in blade pitch, ranging from –2 degrees to +2 degrees. Pitch angle changes affect output on FSIG turbines more than on DFIG turbines. That is because FSIG turbine rotor speeds are normally set within 1 percent of synchronous speed.

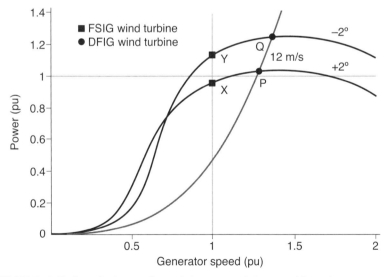

FIGURE 7-4 Pitch angle changes from –2 degrees to +2 degrees with resultant power extraction curves.

Data from Anaya-Lara, O., Jenkins, N., Ekanayake, J., Cartwright, P., & Hughes, M. (2009). *Wind energy generation: Modelling and control.* West Sussex, England: John Wiley & Sons, Ltd.

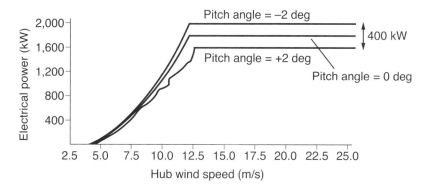

FIGURE 7-5 Generator unloading via pitch angle control methodology.

Data from Anaya-Lara, O., Jenkins, N., Ekanayake, J., Cartwright, P., & Hughes, M. (2009). *Wind energy generation: Modelling and control.* West Sussex, England: John Wiley & Sons, Ltd.

Refer to Figure 7-4 and imagine an FSIG turbine in wind speeds of 12 m/s^{-1}. It operates at point X for 2 degrees of positive pitch. It operates at point Y for 2 degrees of negative pitch. Now think of a DFIG wind turbine in the same wind speed. It employs a power controller to operate for maximum power extraction. Thus, it operates at point P for 2 degrees of positive pitch. It operates at point Q for 2 degrees of negative pitch.

FIGURE 7-5 shows effects of changing pitch angle for wind speeds above and below the power control reference point (P_{ref}). Below rated wind speed, the 4-degree change can offer unloading of up to 400 kW. At winds above ratings, pitch angle is necessarily much higher than 2 degrees.

Power production controllers on variable-pitch turbine blades operate in two regions. See **FIGURE 7-6**. Region A represents operations at above-rated

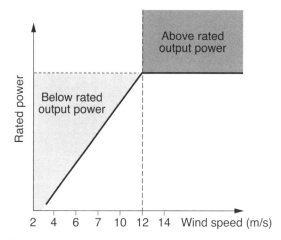

FIGURE 7-6 Pitch angle controller operating regions.

Data from Anaya-Lara, O., Jenkins, N., Ekanayake, J., Cartwright, P., & Hughes, M. (2009). *Wind energy generation: Modelling and control.* West Sussex, England: John Wiley & Sons, Ltd.

power output. Region B represents operations at below-rated power output. FSIG and DFIG turbines operating in region A use the pitch-to-feather power control strategy to sustain rated output. FSIG turbine controllers operate in region B at minimum pitch angle as permitted by control limits. DFIG turbines' power controllers operate constantly based on the preset torque-speed curve. This allows consistent maximum power extraction in region B conditions.

It is possible to use the same kind of pitch angle controller in DFIG and FSIG turbines. Operations in region A are based on the pitch-to-feather method, but it is a modified version. It uses P_{ref} regulation for low- and high-frequency reaction. Operations in region B minimum pitch angle control are governed for frequency reaction, but DFIG turbines use this method in conjunction with power electronic controllers.

Full Rated Power Electronic Interface

Full rated power electronic converters are an important control mechanism. Their interface can connect induction and synchronous generators to a grid. They use passive rectifiers and boost converters. At low speed, those mechanisms increase low-speed operation generator voltage. Generator active power is controllable. A grid inverter controls it by connecting the direct current (DC) link with the network. These mechanisms in the same configuration can also control reactive power. You can see that full rated power electronic converters provide a high degree of generator control. Thus, these turbines are very useful for both primary and secondary response occasional service. See **FIGURE 7-7** for a visual representation of this control configuration.

NOTE

Full rated power electronics are common in small and midsize wind turbines, as building the most sophisticated generators for these smaller designs is cost prohibitive. They are often simply called turbine controllers.

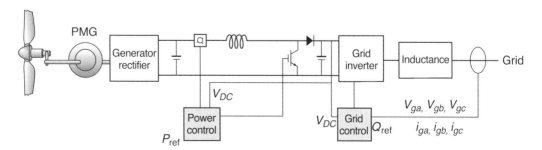

FIGURE 7-7 Full-rated power electronic converter control of reactive and active power for synchronous generators.
Adapted from Blaabjerg, F., & Chen, Z. (2006). *Power electronics for modern wind turbines.* Copenhagen, Denmark: Aalborg University.

Topological Area Control Considerations

Many countries' networks these days include a high dependence on wind farms. This is especially true in Europe. Because of this dependence, modern turbines must meet high technical demands. These demands revolve around frequency control, voltage control, active and reactive power control, and occasional service needs. Overall wind farm performance depends on wind turbine types. It also depends on grid **topology**. Topology is the overall layout of the grid over physical terrain. It takes into account wind patterns and electrical needs based on regional geographical features.

FIGURE 7-8 shows the layout of a wind farm using power electronic converters. These help control active and reactive power. They also allow variable-speed operations. This maximizes wind power extraction at varying wind speeds. They also reduce mechanical stress and noise levels. These systems are good for offshore wind farms.

FIGURE 7-9 illustrates an induction generator-based wind farm. Static synchronous compensators (STATCOMs) provide reactive power control. They also help with voltage control. Lastly, they provide for the reactive power demand in these wind farms.

FIGURE 7-10 shows a wind farm layout where each wind turbine has its own power electronic converter. This allows each turbine to operate at optimal speed for maximum power extraction.

FIGURE 7-11 depicts a wind farm using high-voltage DC (HVDC) transmission. These systems are also good options for offshore wind turbines. This is especially true for offshore systems located far away from land. HVDC systems based on voltage source converters (VSCs) help transmit power over long distances.

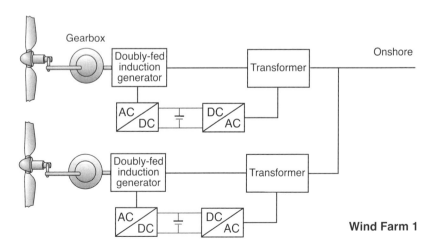

FIGURE 7-8 Spatial diagram of wind farm using power electronic converters.

Adapted from Blaabjerg, F., & Chen, Z. (2006). *Power electronics for modern wind turbines.* Copenhagen, Denmark: Aalborg University.

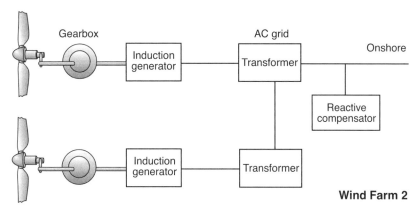

FIGURE 7–9 Spatial diagram of wind farm with induction generators.

Adapted from Blaabjerg, F., & Chen, Z. (2006). *Power electronics for modern wind turbines.* Copenhagen, Denmark: Aalborg University.

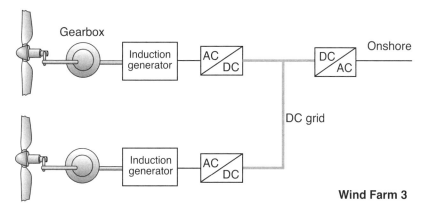

FIGURE 7–10 Spatial diagram of wind farm with individual turbine power electronic converters.

Adapted from Blaabjerg, F., & Chen, Z. (2006). *Power electronics for modern wind turbines.* Copenhagen, Denmark: Aalborg University.

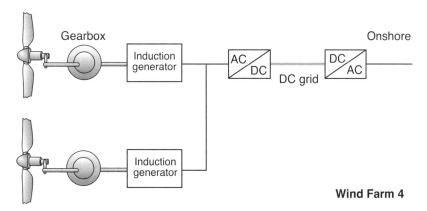

FIGURE 7–11 Spatial diagram of wind farm using high-voltage DC transmission.

Adapted from Blaabjerg, F., & Chen, Z. (2006). *Power electronics for modern wind turbines.* Copenhagen, Denmark: Aalborg University.

TABLE 7-1 COMPARISON OF FOUR WIND FARM TOPOLOGIES

Configuration	Wind Farm 1	Wind Farm 2	Wind Farm 3	Wind Farm 4
Individual speed control	✓		✓	
Control active power electronically	✓		✓	✓
Control reactive power	✓	Centralized	✓	✓
Short circuit (active)	Partly	Partly	✓	✓
Control bandwidth	10–100 ms	200 ms to 2 s	10–100 ms	10 ms to 10 s
Standby function	✓		✓	✓
Soft starter needed		✓		
Rolling capacity on grid	✓	Partly	✓	✓
Redundancy	✓	✓		
Cost	$$	$	$$	$$

Note: $ indicates lower cost, $$ indicates more expensive

Adapted from Blaabjerg, F., & Chen, Z. (2006). *Power electronics for modern wind turbines.* Copenhagen, Denmark: Aalborg University.

When the DC voltage arrives at the land-based station, it is converted back into AC. Then it is fed into the grid. VSC-based HVDC turbines lose less power in transmission than thyristor-based HVDC systems.

TABLE 7-1 compares the various characteristics of the four different wind farm/grid topologies shown in the previous figures. Other options exist. They include field excited synchronous machines or permanent magnet-based machines. These can be used with the wind farms depicted in Figures 7-10 and 7-11. Additionally, multipole generators may not need to use a gearbox for torque/speed conversion.

WHAT DO FISH AND WIND ENERGY HAVE IN COMMON?

Have you ever wondered what wind turbines have in common with fish? Probably not, but John Dabiri has. He is the head of the Biological Propulsion Laboratory at the California Institute of Technology (Caltech). In this role, he focuses on the dynamics of fluids on a daily basis. And because fish live in water, he's spent lots of time observing the way they interact

WHAT DO FISH AND WIND ENERGY HAVE IN COMMON? (Continued)

with their liquid environment. It turns out that fish may be able to teach man a lot about how to maximize wind energy extraction.

In Caltech's *Science Daily*, Dabiri says, "I became inspired by observations of schooling fish, and the suggestion that there is constructive hydrodynamic interference between the wakes of neighboring fish. Many of the same physical principles can be applied to the interaction of vertical axis wind turbines."

Current HAWTs face challenges because they need to be spaced far apart for optimal functionality. This is due to the disturbance that each turbine creates as it extracts power from wind. Previously, VAWTs have not been popular in wind farms because they are not as efficient at extracting wind's energy, but in areas with dense populations, space for large wind farms is usually limited.

So Mr. Dabiri is studying how the use of many more VAWTs in a limited space can produce more power than HAWTs on the same space.

As fish swim through water, they leave vortices behind. They are similar to the vortices that VAWTs create—some turn in a clockwise fashion, while others turn in a counterclockwise fashion. Mr. Dabiri has purchased two acres of land north of Los Angeles to test his hypothesis. He and his students are currently testing different VAWT layouts to optimize energy extraction. In this testing, the focus is double. As at any wind farm, wind is needed to power the VAWTs, but unlike modern wind farms, the vortices can actually *help* improve efficiency because they can spin whichever way the wind is blowing. In other words, where a HAWT would lose efficiency due to vortices, VAWTs can take advantage of vortices to *increase* power extraction.

Windspire Energy, a local VAWT manufacturer, has donated several VAWTs to Mr. Dabiri for his testing. Each turbine is about 30 feet tall and 4 feet wide. They can generate up to 1.2 kW of energy each. They are in a dense series of configurations, allowing for many more turbines per square acre than traditional HAWT-based wind farms permit. "This project is unique in that we are conducting these experiments in real-world conditions," says Mr. Dabiri in *Science Daily*. Usually, this kind of experimentation is done in a laboratory or using computer-generated testing models. "In the future, we hope to transition to power generation experiments in which the generated power can be put to use either locally or via a grid connection."

"This leading-edge project is a great example of how thinking differently can drive a meaningful innovation," says Windspire Energy CEO Walt Borland. "We are very excited to be able to work with Mr. Dabiri and Caltech to better leverage the unique attributes of vertical axis [wind turbine] technology."

So although you might never have considered what fish and wind turbines have in common, someone has. And your future career just might include some of the lessons that Mr. Dabiri is learning now.

CHAPTER SUMMARY

In this chapter, you focused on wind turbine control strategies. Controllers permit wind turbines to operate efficiently. Sometimes there are control *strategies* that do the same thing. Control strategies consist of hardware that depends on manipulation for optimizing power extraction.

This chapter started with a discussion of frequency control. Frequency control is key in electrical systems. That is because an entire grid must maintain a certain frequency to service loads efficiently. The frequency cannot be too high or low. Then you studied active stall wind turbines. These are usually fixed-speed operations. Thus, control methods allow for changes in frequency in spite of these turbines' fixed-speed nature. This is important for active stall wind turbine operation flexibility. Next, you covered variable pitch angle control methods. These methods depend on pitching blades for efficiency. Blades change pitch to respond to frequency concerns. These blades and their controls are another way to allow efficient turbine operations. Last, you studied full rated power electronic converters and then topological wind system considerations. Wind system topology concerns are vital because geographical features affect power generation. They also affect how power is transmitted and distributed. You compared four common topological wind system layouts.

CHAPTER KEY CONCEPTS AND TERMS

Dead band
Droop governors
Occasional service
Pitch angle power control

Slow primary response
Spilling
Topology

CHAPTER ASSESSMENT: WIND TURBINE CONTROL

1. When electrical system load increases, system frequency does which of the following?
   - ❑ **A.** Increases
   - ❑ **B.** Decreases
   - ❑ **C.** Stays the same
   - ❑ **D.** None of the above

2. The normal acceptable limit for system frequency fluctuations is approximately _____ _____

3. Which of the following consists of normal frequency restoration?
   - ❑ **A.** Occasional service
   - ❑ **B.** Primary response
   - ❑ **C.** Secondary response
   - ❑ **D.** None of the above

4. Fixed-speed wind turbines with ratings higher than 1 MW are inertially constrained. The constraints range from _____ to _____ seconds.

5. There is a direct link between rotor speed and system frequency in DFIGs with conventional controls.
   - ❑ **A.** True
   - ❑ **B.** False

6. Direct-drive generators increase the power transfer from rotor to generator.
   - ❑ **A.** True
   - ❑ **B.** False

7. In the variable pitch angle control method, the blade pitch is gradually _____ as wind speed increases.

8. Spilling leaves no margin for extra wind turbine loading.
   - ❑ **A.** True
   - ❑ **B.** False

9. Below rated power, blade pitch is usually fixed for optimal energy extraction.
   - ❑ **A.** True
   - ❑ **B.** False

10. Pitch angle changes affect output on FSIG turbines more than on _____ turbines.

11. Power production controllers on variable-pitch turbine blades operate in how many region(s)?
    - ❑ **A.** One
    - ❑ **B.** Two
    - ❑ **C.** Three
    - ❑ **D.** Four

12. In full rated power electronic interface control systems, what connects the DC link with the network?
    - ❑ **A.** Reactive power converter
    - ❑ **B.** Voltage source converter
    - ❑ **C.** Grid inverter
    - ❑ **D.** Voltage source controller

13. Wind farms that use power electronic converters are especially useful in onshore wind energy systems.
    - ❑ **A.** True
    - ❑ **B.** False

14. _____-based wind energy systems are useful for offshore systems located far from land.

15. Multipole generators may not need to use a _____ for torque/speed conversion.

Rotor Dynamics

IN THIS CHAPTER, you will study the basics of rotor dynamics and blade bending. **Rotor dynamics** is a specialized field of study in wind energy. It is based on mechanical studies of torque and speed as they apply to the rotor and drive shaft. Of course these are affected by blade bending. You should be familiar with the basic principles of blade bending. In this chapter, you will read more about how blade bending contributes to rotor dynamics.

This chapter opens with a review of blade bending concepts. You will read about two different kinds of blade bending. One is called out-of-plane bending and the other is called in-plane bending. You will study what these kinds of bending are and which are important in the field of rotor dynamics. Next, you will study the three-mass model of rotor dynamics study. This includes how the blades and rotor are split into varying sections for this study. You will also explore how to calculate the five parameters of the three-mass model, and you will study a real-life example of how the three-mass model is applied to a particular wind turbine. Then, you will move on to study the two-mass model of rotor dynamics study. This model is a simpler version of rotor dynamics study. You will read about its three parameters and how to calculate them. You will study the application of this model to a real-life wind turbine. This chapter closes with a brief overview of turbine assessment. You will review some visuals and compare and contrast the results between single-mass, two-mass, and three-mass model studies.

Chapter Topics

This chapter covers the following topics and concepts:

- Basics of blade bending concepts
- Three-mass model of rotor dynamics
- Two-mass model of rotor dynamics
- Basics of FSIG assessment using single-, two-, and three-mass models of rotor dynamics

Chapter Goals

When you complete this chapter, you will be able to:

- Describe the basic concepts of blade bending, including in-plane and out-of-plane bending
- Explain the basis of the three-mass model of rotor dynamics
- Calculate the five parameters of the three-mass model of rotor dynamics
- Chart a real-life example of the three-mass model of rotor dynamics, given three of the necessary five parameters
- Explain the basis of the two-mass model of rotor dynamics
- Calculate the three parameters of the two-mass model of rotor dynamics
- Chart a real-life example of the two-mass model of rotor dynamics, given two of the necessary three parameters
- Compare and contrast the single-, two-, and three-mass model results of FSIG rotor dynamics analyses graphs

Blade Bending

Two main factors contribute to turbine torque oscillations. The first main factor is the shaft's torsional flexibility. The second is the blades' bending flexibility. Additionally, individual blade characteristics can affect oscillations. For example, each blade has slight variations in mass distribution. They also vary a little in stiffness and twist angle, but for your current studies, you can assume equal characteristics when studying rotor dynamics.

There are two spatial planes to consider when studying blade bending. The first is called **out-of-plane bending**. Out-of-plane bending refers to the blade bending perpendicular to the rotor plane. The second is called **in-plane bending**. In-plane bending refers to the blade motion on the rotor plane. Out-of-plane bending is normal within the direction of rotor rotation. Thus, it is not linked directly to the drive train. For this reason, out-of-plane bending does not figure into drive-train mechanics. Some in-plane bending characteristics do count when considering drive-train bending.

In-plane bending has two mode categories: symmetric and asymmetric. Two asymmetric modes do not couple with the drive train, so you can disregard them for now. The symmetric characteristics do affect rotor dynamics. You will read more about these in the next section.

Three–Mass Model

See **FIGURE 8-1**. O represents the rotor hub. The blades are shown in typical bending modes. The darker sections near the rotor are labeled $A1$, $A2$, and $A3$. These sections represent the stiffest part of each blade. The stiff parts of the blade have little to no bending modes. They have a hub inertia represented by the variable J_2 in Figure 8-1B. The lighter sections extending out from there are the bending sections. They are labeled $B1$, $B2$, and $B3$. These sections are more susceptible to bending. That is because they are necessarily more flexible than the A sections. The *combined A1B1*, *A2B2*, and *A3B3* sections carry a circumference-based inertia of J_2 in Figure 8-1B. J_1 and J_2 connect via the springs in Figure 8-1B. These springs effectively represent the individual blades' total flexibility.

FIGURE 8-2 is a detailed expansion of Figure 8-1B. J_1 represents the flexible blade section inertia. J_2 equals the combined hub and rigid blade section inertia. J_3

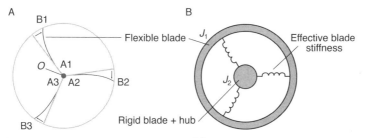

FIGURE 8-1 In-blade symmetric bending modes (A) and equivalent torsional force diagram (B).

Adapted from Anaya-Lara, O., Jenkins, N., Ekanayake, J., Cartwright, P., & Hughes, M. (2009). *Wind energy generation: Modelling and control.* West Sussex, England: John Wiley & Sons, Ltd.

FIGURE 8-2 Three-mass drive-train model with blade and drive shaft flexibility values.

Adapted from Anaya-Lara, O., Jenkins, N., Ekanayake, J., Cartwright, P., & Hughes, M. (2009). *Wind energy generation: Modelling and control.* West Sussex, England: John Wiley & Sons, Ltd.

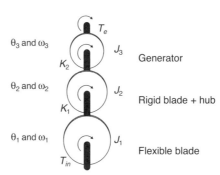

represents generator inertia. K_1 is blade stiffness, and K_2 is low- and high-speed shaft stiffness. J_3, K_2, and (J_1+J_2) are known variables. The three-mass model consists of five parameters. Thus, you need two more equations to figure out the total dynamic values in the three-mass model.

FIGURE 8-3 is a schematic representation of the three-mass drive-train model. The following equations are all the equations needed to calculate dynamics, vibration frequencies, and oscillation magnitudes.

Three-Mass Model Dynamic Equations

1. $J_1 \dfrac{d^2}{dt^2}\theta_1 = -K_1(\theta_1 - \theta_2)$

2. $J_2 \dfrac{d^2}{dt^2}\theta_2 = -K_2(\theta_2 - \theta_1) - K_2(\theta_2 - \theta_3)$

3. $J_3 \dfrac{d^2}{dt^2}\theta_3 = -K_2(\theta_3 - \theta_2)$

Three-Mass Model Vibration Frequency Equations

1. $\displaystyle \int_1 = \frac{1}{2\pi}\left(-\frac{b}{2} - \frac{\sqrt{b^2 - 4c}}{2}\right)^{\frac{1}{2}}$

2. $\displaystyle \int_2 = \frac{1}{2\pi}\left(-\frac{b}{2} + \frac{\sqrt{b^2 - 4c}}{2}\right)^{\frac{1}{2}}$

Three-Mass Model Oscillation Magnitude Equations (Additional Equations)

1. $\dfrac{\theta_1}{\theta_2} = \dfrac{K_1}{(K_1 - J_1\omega^2)}$

2. $\dfrac{\theta_2}{\theta_3} = \dfrac{(K_2 - J_3\omega^2)}{K_2}$

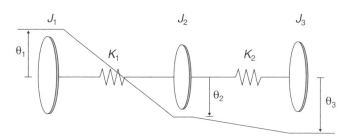

FIGURE 8-3 Schematic diagram of the three-mass drive-train model.

Adapted from Anaya-Lara, O., Jenkins, N., Ekanayake, J., Cartwright, P., & Hughes, M. (2009). *Wind energy generation: Modelling and control.* West Sussex, England: John Wiley & Sons, Ltd.

Three-Mass Model Operational Example

In this example, you should consider a 300 kW fixed-speed induction generator (FSIG) wind turbine. Here are the known three-mass model parameters:

1. $J_3 = 0.102 \times 10^6 \, kgm^2$
2. $J_1 + J_3 = 0.129 \times 10^6 \, kgm^2$
3. $K_2 = 5.6 \times 10^7 \, Nmrad^{-1}$

These parameters all refer to the low-speed, high-torque shaft. To locate the rotor vibration frequencies, the rotor is excited. Application of 20 percent voltage sag to the generator at 5 s for 200 ms accomplishes the excitation. This leaves 80 percent retained voltage in the generator.

FIGURE 8-4A represents the low-speed shaft torque after applying the 20 percent voltage sag. **FIGURE 8-4B** represents the matching frequency spectrum. You can substitute the spectral frequencies from Figure 8-4B in the vibration frequency equations. If you do, you will be able to calculate all five three-mass model parameters:

1. $J_1 = 0.111 \times 10^6$
2. $J_2 = 0.018 \times 10^6$
3. $J_3 = 0.102 \times 10^6 \, kgm^2$
4. $K_1 = 2.1 \times 10^7$
5. $K_2 = 5.6 \times 10^7$

Now you can start to draw conclusions. For example, K_1 (effective blade flexibility) is 0.4 times smaller than K_2 (shaft flexibility). Thus, blade flexibility

> **NOTE**
>
> Computer-based modeling of generators and turbine fluid dynamics is evolving quickly as new interest in wind power develops around the world. Specialized software, as well as modules available for general purpose finite element analysis and multi-physics software, is now available to help designers, particularly for HAWTs. The field is still very new for modeling VAWTs.

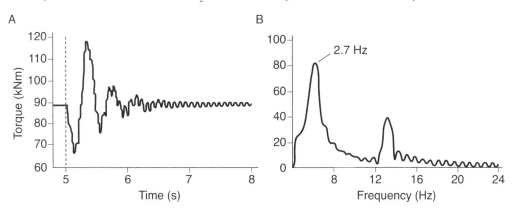

FIGURE 8-4 20 percent voltage sag low-speed shaft torque and harmonic spectrum (A) and low-speed shaft torque FFT (B).

Data from Anaya-Lara, O., Jenkins, N., Ekanayake, J., Cartwright, P., & Hughes, M. (2009). *Wind energy generation: Modelling and control*. West Sussex, England: John Wiley & Sons, Ltd.

is much more important than shaft flexibility, but the blade *and* shaft flexibility representations increase the model's order. This might not always be necessary for large power system studies. Also, dynamic studies need to consider the most important frequencies. They are key in equipment with the lowest operating frequencies. Therefore, you will now read about the two-mass model.

N anotechnology is being used to develop new materials that provide better blade dynamic properties. Nano-Clay, already used in automotive bumpers, is being used to strengthen blades to provide optimal bending properties in order to meet engineering goals.

Two-Mass Model

The two-mass model considers shaft and blade flexibility, but it only considers the dominant low-frequency component of a turbine. Figure 8-2 shows two coupled modes ($\int_1$ and $\int_2$). Now ignore blade flexibility ($K_1 = \infty$). That drives the two-mass model to have a natural vibration frequency of $\int_{shaft}$. Consider exclusion of shaft flexibility ($K_2 = \infty$). In that case, the two-mass model would have a vibration frequency of $\int_{blade}$. See **FIGURE 8-5** for a schematic representation of

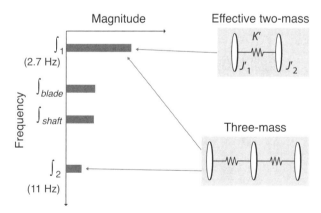

FIGURE 8-5 Multimass system model frequency components. $\int_1$ and $\int_2$ represent three-mass model coupled modes' frequency components. $\int_{shaft}$ represents shaft flexibility in the two-mass model. $\int_{blade}$ represents blade flexibility in the two-mass model.

Adapted from Anaya-Lara, O., Jenkins, N., Ekanayake, J., Cartwright, P., & Hughes, M. (2009). *Wind energy generation: Modelling and control.* West Sussex, England: John Wiley & Sons, Ltd.

these situations. Neither situation shows the true rotor dynamics' dominant frequency vibrations. That is because the three-mass model accounts for these values.

Here are the relevant formulas for determining the three parameters of the two-mass model of rotor dynamics:

Two-Mass Model Dynamic Equations

1. $J_1 \dfrac{d^2}{dt^2} \theta_1 = -K_1(\theta_1 - \theta_2)$

2. $J_2 \dfrac{d^2}{dt^2} \theta_2 = -K_2(\theta_2 - \theta_1)$

Two-Mass Model Vibration Frequency Equations

1. $\displaystyle \smallint_1 = \frac{1}{2\pi} \sqrt{\frac{K}{\left(\dfrac{1}{J_1} + \dfrac{1}{J_2} \right)^{-1}}}$

Two-Mass Model Oscillation Magnitude Equation (Additional Equation)

1. $\dfrac{\theta_1}{\theta_2} = -\dfrac{J_1}{J_2}$

Two-Mass Model Operational Example

In this example, you should again consider a 300 kW FSIG wind turbine. You need three equations to calculate the two-mass model parameters. The parameters are J'_1, J'_2, and K'. They are calculated in the following manner:

1. The turbine inertia's total moment is stated as:

$$J'_1 + J'_2 = J_{total} = 0.231 \times 10^6 \, kgm^2$$

2. The rotor's lowest-frequency component was 2.7 Hz. This enters into the preceding natural vibration frequency formula in the following manner:

$$\smallint_1 = \frac{1}{2\pi} \sqrt{\frac{K'}{\left(\dfrac{1}{J'_1} + \dfrac{1}{J'_2} \right)^{-1}}} = 2.7$$

3. The last equation considers the effective two- and three-mass models' oscillation magnitude ratios. They are used in the following way:

$$\frac{\theta'_1}{\theta'_2} = -\frac{J'_1}{J'_2} = \frac{\theta_1}{\theta_3} = \frac{K_1 \left(K_2 - J_3 \omega^2 \right)}{K_2 \left(K_1 - J_1 \omega^2 \right)} = -0.92$$

If you work through the previous three equations, you will see the three parameters' values. Here they are for reference:

1. $J'_1 = 0.111 \times 10^6$
2. $J'_2 = 0.12 \times 10^6$
3. $K' = 1.66 \times 10^7$

The conclusion is that both full rotor dynamics and the two-mass model both yield approximately the same response results.

FSIG Turbine Performance Assessment

To test FSIG performance, you would apply an electrical fault at 5 s for 200 ms. This leads to a terminal voltage drop. The drop is by 20 percent, retaining 80 percent power in the terminals.

Once again the full rotor dynamics response closely matches that of the two-mass model, but the two-mass model based only on shaft flexibility does not match up with actual turbine response.

OFFSHORE WIND TECHNOLOGY

The US Department of Energy (DoE) has an Offshore Wind Technology Program. This program has two main goals. The first is reduction in offshore-produced energy costs. The second is reduction of offshore wind system deployment timelines. By 2030, this program hopes to help develop 54 gigawatts of offshore capacity in the United States. The DoE anticipates that production and delivery costs will be between $0.07 and $0.09 per kilowatt-hour.

Wind resources at sea are much greater than on land. In addition, transport considerations are easy to determine because offshore systems require no roads. For these reasons, offshore wind turbines can be much larger than onshore turbines, but offshore marine conditions are much harsher than those on land. So turbines have to be much sturdier in their construction. Also, deep-water installations require new platform and foundation schemes, as shown in **FIGURE 8-6**. These include submersible foundations and even floating platforms!

The DoE program is also working to remove market-entry barriers for offshore wind system deployment. These barriers include the following items, among others:

- Wind resource planning activities
- Siting and regulatory hurdles, especially focused on resolving environmental concerns
- Complementary installation issues, revolving around transport and installation particularities

Once the DoE program is at a more advanced stage, it will enter partnerships. The groups it will partner with include commercial wind power developers, research groups, and utilities.

OFFSHORE WIND TECHNOLOGY (Continued)

The Department of Energy is also working hard to coordinate information sharing between several other federal governmental entities. These include the Department of the Interior, the National Oceanic and Atmospheric Administration, the Department of Defense, the Environmental Protection Agency, the US Fish and Wildlife service, and others.

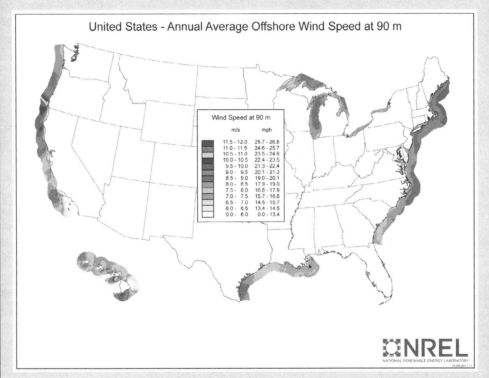

FIGURE 8-6 US Department of Energy diagram of offshore wind system characteristics currently under study.
Courtesy of the Office of Energy Efficiency and Renewable Energy (EERE)

CHAPTER SUMMARY

In this chapter, you have studied the basics of rotor dynamics and blade bending. Rotor dynamics is a specialized field of study in wind energy. It is based on mechanical studies of torque and speed as they apply to the rotor and drive shaft. Of course these are affected by blade bending. Before reading this chapter, you were already familiar with the basic principles of blade bending. In this chapter, you read about how blade bending contributes to rotor dynamics.

This chapter started with a review of blade bending concepts. You read about two different kinds of blade bending. One is called out-of-plane bending and the other is called in-plane bending. You then studied what these kinds of bending are and which are important in the field of rotor dynamics. Next, you covered the three-mass model of rotor dynamics study. This includes how the blades and rotor are split into varying sections for this study. You also studied how to calculate the five parameters of the three-mass model. And you studied a real-life example of how the three-mass model is applied to a particular wind turbine. Then you moved on to study the two-mass model of rotor dynamics study. This model is a simpler version of rotor dynamics study. You read about its three parameters and how to calculate them. Again, you studied the application of this model to a real-life wind turbine. This chapter closed with a brief overview of turbine assessment. You reviewed some visuals and compared and contrasted the results between single-mass, two-mass, and three-mass model studies.

CHAPTER KEY CONCEPTS AND TERMS

In-plane bending
Out-of-plane bending
Rotor dynamics

CHAPTER ASSESSMENT: ROTOR DYNAMICS

1. Which of the following is the best definition of *rotor dynamics*?
   - ❑ A. A specialized field of study in wind energy. It is based on electrical studies of torque and speed as they apply to the rotor and drive shaft.
   - ❑ B. A specialized field of study in wind energy. It is based on mechanical studies of voltage and speed as they apply to the rotor and drive shaft.
   - ❑ C. A specialized field of study in wind energy. It is based on mechanical studies of torque and speed as they apply to the rotor and drive shaft.
   - ❑ D. A specialized field of study in wind energy. It is based on mechanical studies of torque and speed as they apply to the gearbox and drive shaft.

2. The symmetric characteristics of _____ blade bending must be taken into account in the study of rotor dynamics.

3. Out-of-plane bending does not need to be considered in the study of rotor dynamics.
   - ❑ **A.** True
   - ❑ **B.** False

4. In this chapter, J_2 is taken to stand for _____ _____ _____.

5. Which of the following variables are known in the three-mass model of rotor dynamics?
   - ❑ **A.** J_3, K_2, and (J_1+J_2)
   - ❑ **B.** J_1, J_3, and K_2
   - ❑ **C.** J_3, K_1, and K_2
   - ❑ **D.** None of the above

6. The following equation is one of the equations for calculating which of the following options?

$$J_1\frac{d^2}{dt^2}\theta_1 = -K_1(\theta_1 - \theta_2)$$

   - ❑ **A.** Three-mass vibration frequency
   - ❑ **B.** Three-mass oscillation magnitude
   - ❑ **C.** Three-mass dynamic model performance
   - ❑ **D.** None of the above

7. The following equation is used to calculate which of the following items?

$$\int_2 = \frac{1}{2\pi}\left(-\frac{b}{2} + \frac{\sqrt{b^2 - 4c}}{2}\right)^{\frac{1}{2}}$$

   - ❑ **A.** Three-mass vibration frequency
   - ❑ **B.** Three-mass oscillation magnitude
   - ❑ **C.** Three-mass dynamic model performance
   - ❑ **D.** None of the above

8. Using the three-mass model example given in this chapter, you learned that shaft flexibility is much more important than blade flexibility in the study of rotor dynamics.
   - ❑ **A.** True
   - ❑ **B.** False

9. The two-mass model of rotor dynamics takes into account only the highest-frequency components of a wind turbine.
   - ❑ **A.** True
   - ❑ **B.** False

10. Which of the following variables is used to denote blade flexibility in this chapter?
   ❏ A. J_1
   ❏ B. J_2
   ❏ C. K_1
   ❏ D. K_2

11. What does the following equation allow you to calculate?

$$J_2 \frac{d^2}{dt^2}\theta_2 = -K_2(\theta_2 - \theta_1)$$

   ❏ A. Two-mass dynamic model performance
   ❏ B. Two-mass vibration frequency
   ❏ C. Two-mass oscillation magnitude
   ❏ D. None of the above

12. What does the following equation allow you to calculate?

$$\frac{\theta_1}{\theta_2} = -\frac{J_1}{J_2}$$

   ❏ A. Two-mass dynamic model performance
   ❏ B. Two-mass vibration frequency
   ❏ C. Two-mass oscillation magnitude
   ❏ D. None of the above

13. Oscillation magnitude is a consideration of two- and three-mass models of rotor dynamics.
   ❏ A. True
   ❏ B. False

14. Full rotor dynamics and the _____ _____ both yield about the same results as each other.

15. To test FSIG rotor dynamic performance, you need to apply an electrical fault at _____ s for _____ ms.

Wind Farms

IN THIS CHAPTER, you will study some of the concerns with wind farms and power distribution. The chapter opens by studying how wind farms affect the dynamics of networks. You will first look at how fixed-speed induction generator (FSIG) turbines fit into a network scheme regarding network damping. Then you will review how doubly fed induction generator (DFIG) turbines integrate into a grid. You will read about the similarities and differences between these generator options. Specifically, you will study their impact on grid power and voltage control. The next section covers some specific issues with connecting wind turbines to large networks. You will read about issues like frequency, active power control, short-circuit voltage, reactive power control, voltage flicker, harmonics, and network stability. You will also read about how utilities and regulators deal with problems in these areas of grid integration. The chapter ends by covering substations. You will read about the main types of substations and their functions. You will also review the equipment that each kind of substation has to complete its functions.

Chapter Topics

This chapter covers the following topics and concepts:

- The influence of wind farms on network dynamic performance, including the effects of FSIG and DFIG network integration
- Wind farm grid connection concerns, including frequency and active power control, short-circuit voltage, reactive power control, voltage flicker, harmonics, and network stability
- Main electrical substation types and purposes, including principal electrical substation equipment and purposes

Chapter Goals

When you complete this chapter, you will be able to:

- Describe the two principal mechanisms by which FSIG grid integration affects network damping
- Explain the two principal mechanisms by which DFIG grid integration affects network damping
- Dialogue on common wind farm integration concerns, including frequency and active power control, short-circuit voltage, reactive power control, voltage flicker, harmonics, and network stability
- Explain how utilities and government entities regulate problems with frequency and active power control, short-circuit voltage, reactive power control, voltage flicker, harmonics, and network stability to maintain network power flow and voltage at optimal levels
- Compare and contrast the functionalities of transmission substations, distribution substations, collector substations, and switching substations
- List and explain the uses of common substation equipment, including transformers, switches, power controllers and converters, circuit breakers, fuses, bus bars, and grounding equipment

Influence of Wind Farms on Network Dynamic Performance

When a wind farm is connected to a larger grid, some unique dynamics come into play. The most important consideration is the wind farm's influence on network damping. The following sections provide an overview of how FSIG and DFIG wind farms affect overall network damping.

FSIG Network Damping

Torque synchronization and damping can illustrate FSIG mechanisms related to network damping.

NOTE

A **bus bar** is a heavy-duty conducting rod. A bus bar carries electrical voltages between loads and supply systems. Bus bars often form part of transformer station equipment for electricity distribution.

You will now review an analysis of this damping contribution. It is based on **FIGURE 9-1**. Network bus bar voltages are defined on two parameters. The first parameter is magnitude, and the second is phase angle.

To simplify this model, you can assume that bus voltages are constant. This means that line-system power variations depend only on voltage phase changes. You can also assume that in oscillatory conditions, the terminal 1 voltage phase

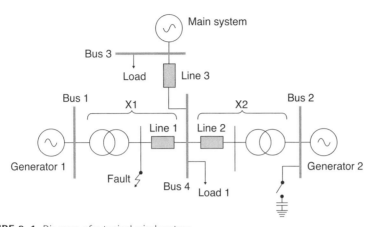

FIGURE 9-1 Diagram of a typical wind system.

Adapted from Anaya-Lara, O., Jenkins, N., Ekanayake, J., Cartwright, P., & Hughes, M. (2009). *Wind energy generation: Modelling and control.* West Sussex, England: John Wiley & Sons, Ltd.

change depends solely on rotor speed. You should also assume that each generator's output is constant.

Increasing generator 1's rotor angle increases its phase angle correspondingly. Then, the increased power flow in line 1 will generate an increase in line 3's power flow. This increases central bus bar voltage phase angle as related to the main system bus bar.

Based on that increase, there is a reduction in phase angle between generator 2's terminal voltage and the central bus bar. Consequently, line 2's power flow is reduced, and that reduces generator 2's output. Because turbine power stays constant, rotor speed then decreases. The change in generator speed reduces slip value. This terminates with a further change in generator 2's power output. Slip always varies when stator supply frequency changes. Additionally, terminal voltage phase changes affect stator frequency. Thus, you can derive that terminal voltage phase changes also affect slip via stator frequency changes.

DFIG Network Damping

Torque synchronization and damping also explain how DFIGs influence system damping. In this case, you should consider a generic DFIG using an FMAC control system. FMAC is short for "flux magnitude and angle control." You should assume in this model that bus bar voltages are constant. Thus, transmission power line voltages vary only based on phase angle variations. You can see an example of this voltage variation in FIGURE 9-2 .

DFIGs respond to slip variations similar to FSIG responses, but in DFIGs, slip variations have much less influence over power variations. This is true because the constant, K_s, is much smaller in DFIGs. The constant relates power output changes to slip changes. For example, a DFIG with an operating slip value of −0.1 will result in a constant value of approximately 10. Compare that

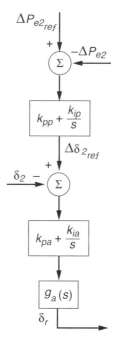

FIGURE 9-2 Power control loop for flux magnitude and angle control method.
Adapted from Anaya-Lara, O., Jenkins, N., Ekanayake, J., Cartwright, P., & Hughes, M. (2009). *Wind energy generation: Modelling and control*. West Sussex, England: John Wiley & Sons, Ltd.

with an FSIG slip value of –0.01. That FSIG will have a constant value of approximately 100. DFIGs' power output varies much more based on controller power loop oscillations.

DFIG power output and voltage control are via manipulation of rotor voltage magnitude and phase angle. So, rotor voltage influences drive voltage changes related to transient reactance. Transient reactance originates in the stator. It influences DFIG power output and terminal voltage. Transient reactance reacts very quickly to rotor voltage changes, and the two values relate directly to each other. Thus, in this model of system damping, you can assume that rotor voltage phase angle changes produce immediate changes in the transient voltage's phase angle.

Connecting Into the Grid

When wind farms link to a grid, the grid's operations may change greatly. In the past, wind turbines couldn't participate in frequency and voltage control, but nowadays, they can. Modern wind farms must meet large-scale performance limits. These limits are based on grid codes. Typical grid codes deal with frequency and voltage control issues. This is done via changes to active and reactive power supply to the grid. Other grid codes deal with wind farm power control regulation. Most grid codes are fulfilled by power electronic converter methods.

Frequency and Active Power

Most electrical generation and distribution systems are alternating current (AC) systems. These normally operate at 50 or 60 Hz. System frequency is directly related to a wind farm's synchronous generator rotating speeds. This indicates that all the generators in an AC wind farm are synchronized, and they run at the same speed. Basically, increased load decreases generator frequency and speed. So frequency control operates to increase (or reduce) generator speed and frequency based on load, but wind production faces a unique challenge. It only produces electricity when wind blows. In a small wind farm connected to a large grid, this might not have much impact, but in large wind farms or small ones connected to small grids, it can have a large effect on delivery. To accommodate frequency increases, turbines sometimes operate at suboptimal levels. This results in waste, due to low usage of wind's energy, but researchers are developing new technologies to compensate for the waste. One example is energy storage technology. Batteries, pump storage, and fuel cells are all undergoing studies for harnessing excess energy. This energy is then used when loads rise.

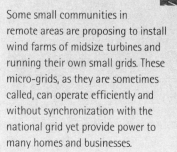

NOTE

Some small communities in remote areas are proposing to install wind farms of midsize turbines and running their own small grids. These micro-grids, as they are sometimes called, can operate efficiently and without synchronization with the national grid yet provide power to many homes and businesses.

Short-Circuit Voltage Levels and Variations

Short-circuit levels in any network are not directly based on voltage quality, but these levels do provide a measure of network strength. Grid disturbance absorption is directly related to network short-circuit limits. If voltage fluctuates at any point in a network, the fluctuation may not affect distant points in the grid. To study this, consider the following:

- Z_k equals the impedance between any given point and a remote grid location.
- U_k equals the nominal voltage of the remote location.
- S_k equals the short-circuit level as measured in **mega volt–amperes**. Mega volt-ampere is the measurement unit for transformer capacity.

The equation for finding S_k follows:

$$\textbf{Equation 1: } S_k = U_k^2/Z_k$$

If impedance is small, voltage variations are also small. This results in a strong grid. If impedance is large, voltage variations are large. Then the grid is considered weak, as shown in FIGURE 9-3.

Reactive Power Control

Reactive power is based on energy oscillations. Specifically, the oscillations relate to stored energy in a grid's capacitive and inductive elements. Synchronous generators produce *and* consume reactive power. This is done via manipulation of

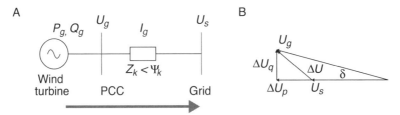

FIGURE 9-3 Simple system with equivalent turbine generator connected; (A) is the system circuit and (B) is the phasor diagram.

Adapted from Blaabjerg, F., & Chen, Z. (2006). *Power electronics for modern wind turbines.* Copenhagen, Denmark: Aalborg University.

magnetization levels. So a high level of magnetization results in high reactive power production and voltage. The opposite is also true.

Reactive power flow currents contribute to voltage and power drops in grids. Large reactive power flow currents can destabilize a network for this reason, but induction generators are good consumers of reactive power. So if they are used in wind energy production systems, they compensate for reactive power current flow problems. They help stabilize system voltages. They also reduce power losses. Turbines with pulse-width modulation (PWM) controllers can use the converters to control reactive power. Consider a PWM-based turbine with a power factor of 1.00. Even this generator can generate or consume reactive power. So it is very helpful as a network voltage control option.

Voltage Flicker

When wind power generation fluctuates, visible light fluctuations can occur in the grid. These fluctuations are called **voltage flicker**. Sometimes this is just called flicker. Generator switching and capacitor switching lead to generator output voltage changes. Sometimes these changes cause voltage flicker. Utilities establish regulations on allowable voltage flicker. So power suppliers must comply with these guidelines. They must ensure their generation equipment causes no voltage flicker past acceptable limits.

The International Electrotechnical Commission (IEC) publication 1000-3-7 has established guidelines for allowable flicker. These guidelines relate to medium- and high-voltage networks. Flicker levels are calculated in short-term severity. This is measured over a 10-minute period. P_{st} is the variable for flicker severity. The lowest level of P_{st} is 1. This level indicates a barely visible flicker level. Flicker values increase from there, based on flicker severity. P_{lt} equals flicker levels as measured over 2-hour periods. This is for defining long-term flicker rates. IEC publication 61000-4-15 guides **flickermeter** design. Flickermeters are devices that measure light waves in the spectrum visible to the human eye. They help establish P_{st} and P_{lt} values in a network.

NOTE

We have all seen the lights flicker during electrical and wind storms, but did you know that flicker happens for other reasons than power lines being disturbed? Flicker can occur when power production facilities come online to the grid or shut down. You may not see this in your incandescent lamp, but you might in a CFL lamp or LED lamp. Even smaller voltage flicker may not be perceptible in lights but may affect sensitive power equipment such as computers and motors.

Harmonics

Harmonics are a distortive characteristic in voltage and current waveforms. All **periodical functions** can be expressed as a totality of sinusoidal waveforms with varying frequencies. This includes the fundamental frequency. It also includes a series of fundamental component integer multiples.

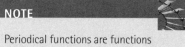

NOTE

Periodical functions are functions that repeat over and over. They can be expressed in mathematical terms.

Depending on harmonic magnitudes, harmonics can damage electronic equipment. All harmonics augment currents. They also cause capacitor overheating. That is because capacitor impedance decreases as harmonic distortion increases. In analog telephone networks, high harmonics cause sound distortion or noise in the line. A piece of equipment can have several parts that emit harmonics. Thus harmonics are measured as a machine's **total harmonic distortion (THD)**. This is the sum of all harmonics emitted by a machine's components.

Wind turbine harmonics control is normally via pulse-width switching converters. They have switching frequencies of around a few kHz. Higher-frequency harmonics are minimal in nature. Usually harmonic filters eliminate or control their effects on equipment.

Network Stability

Network stability issues revolve around network faults. The three main network faults causing instability are as follows:

- Transmission line tripping—Transmission line tripping is also called overload. Overload or component failure disrupts active and reactive power flow. Even when operating generator capacity is sufficient, large voltage drops can occur. This is because reactive power has to flow through new paths in the grid. This drives voltage in down line grid points below stable parameters. Often, a low-voltage period is followed by total power loss in the down line grid points.

- Loss of generation capacity—Loss of generation capacity begins large, temporary power imbalances. If remaining generators have enough spin reserve, they increase frequency to compensate, but if remaining generators are running at full capacity, the power imbalance continues. Then the grid may suffer complete power loss. The most common fix for this is disconnecting power to part of the network. This continues until full generation capacity resumes. This reduces the number of affected consumers in a power loss situation.

- Short circuits—Short circuits can take a variety of forms. One common form is a single-phase earth fault. Trees or other line disturbances cause these short circuits. Another common example is a triple-phase short circuit. This is defined as low impedance in the short-circuit path. Often, a relay protection system in the lines disconnects and recloses quickly. That eliminates the short circuit. Sometimes affected equipment automatically disconnects after a few hundred milliseconds. In all cases, there is a short low- or no-voltage period before transmission normalizes. If a wind farm with no control mechanisms is part of the grid, it may automatically disconnect and worsen the stability situation. Thus, many utilities require that wind farms contain ride-through control systems. This avoids making a low-voltage system worse.

Power System Substations

Electrical substations are network locations, each with a special purpose. They are where voltage levels get stepped up or stepped down. This is done to deliver the correct voltage to transmission systems. Substations also transfer power among and between grid points. They also provide backup and short-circuit protection. There are several different kinds of substations. Often voltage flows through more than one substation before delivery. Substations usually contain equipment for switching, transmission control, and transformers. Large substations contain circuit breakers. They override short circuits or system overloads. Small substations contain recloser circuit breakers or fuses for the same purpose. Many substations contain capacitors and voltage regulators.

NOTE

Distribution substations move a great deal of power. In the United States, most distribution stations manage at least 5 megawatts of electricity.

Some substations are in surface-based fenced enclosures. These are properly grounded. This avoids damage to people and property located nearby. There are also underground locations. Some substations are located inside buildings. Those buildings are specially made. Sometimes skyscrapers contain more than one substation. These are usually to avoid ambient noise pollution. They also protect substation elements from harsh weather events that can damage them.

Types of Substations

There are several different kinds of electrical substations. Here you will read more about transmission, distribution, collector, current changing, and switching substations.

Transmission Substations

Transmission substations connect two or more transmission lines. A simple connection has all lines carrying equal voltage levels. These connect to high-voltage switches. They direct power from one line to another. In many cases, connected lines' voltage must undergo transformation. In these situations, usually there are also capacitors, reactors, or compensators. These help control appropriate power flow between lines. Simple transmission systems may have just a bus and some circuit breakers. Many large transmission substations cover many acres. They usually contain many voltage and current transformers, relays, and supervisory control and data acquisition (SCADA) systems.

Distribution Substations

Distribution substations reduce voltage levels before delivery to most consumers. Unless a consumer uses large amounts of high-voltage electricity, it is not economical or safe to connect to the high-voltage transmission system. A distribution substation's inputs come from several transmission lines. The output consists of several feeders. These lead to various grid points. Sometimes power flows through several distribution substations before voltage is reduced to the consumer load level. Distribution substations also isolate faults in the grid. This helps utilities control which customer groups suffer from low- or no-voltage situations.

In large cities, complicated distribution substations contain transformers, high-voltage switching equipment, and backup systems on the low-voltage side. Smaller distribution substations contain switches, a few transformers, and minimal control equipment on the low-voltage side.

Collector Substations

A **collector substation** is similar to a distribution substation, but is usually located *within* a wind energy production system. Collector substations mostly have the same equipment as distribution substations, but their job is the opposite. They "collect" the power generated by individual turbines. Then they step it up for high-voltage grid transmission. Some collector substations contain power factor correction equipment. Some also contain high-voltage direct current (HVDC) plants.

Current–Changing and Switching Substations

Current-changing substations contain just HVDC converter equipment. These normally connect offshore wind farms with onshore grids. This is because offshore systems produce direct current (DC) currents. They must be converted before transmission and consumption.

Switching substations do not contain transformers. They only drive power flows between different points within or between grids. Sometimes they contain collection and distribution equipment, but these components are for switching to backup lines or parallelizing circuits. This is helpful when generators fail.

WHERE ARE THE WIND FARMS?

The United States is now the world's top producer of wind-based electricity. Germany and China are a close second and third. Germany has more installed wind turbines. It also has a much smaller land area. So how does the United States produce more electricity? Higher average wind speeds! The US Great Plains region has the most installed generation capacity.

(Continues)

WHERE ARE THE WIND FARMS? (Continued)

This area includes parts of Texas, New Mexico, Colorado, Oklahoma, Kansas, Nebraska, Wyoming, Montana, and North and South Dakota. Texas, Iowa, California, Oregon, and Washington have the highest wind energy production rates. In fact, during the spring of 2011, the Bonneville Power area (Oregon and Washington) produced so much wind energy, along with greater than usual hydropower, that officials calculated that the region's electricity was 100 percent from renewable energy. What's more, the region produced so much electricity that it reached the limit of how much power it could transport to California, Mexico, and Canada. The transmission lines couldn't carry any more.

Massachusetts and Rhode Island have the most offshore wind energy production. The amounts are still very small, however, and they come from turbines too small to qualify as a real "wind farm."

How do utilities decide where to put wind farms? There is wind almost everywhere, but there are other issues with regard to siting. An area with too much wind is not useful. Nor is one with rapidly varying winds over long time periods. Densely populated areas do not have enough room for wind farms, but remote areas can be uneconomical. This is due to long-distance transmission costs. Most state and national parks are off-limits. Many wildlife and nature reserves also cannot host wind farms. An area with good winds might not be usable due to landforms that disrupt wind flow. Efficient energy production requires "smooth" winds.

Though there are many concerns in wind farm siting, capacity keeps growing. That is because technology advances and distribution economics continue evolving. The following links provide information about where US wind turbines are located, areas of high wind energy growth, and some regulations regarding wind farm siting:

www.nationalatlas.gov/articles/people/a_energy.html#two

http://weblink.ci.kenai.ak.us/WebLink8/DocView.aspx?id=48396&page=1&&dbid=0

http://en.wikipedia.org/wiki/Wind_power_in_the_United_States

CHAPTER SUMMARY

In this chapter, you studied some of the concerns with wind farms and power distribution. This chapter opened by studying how wind farms affect networks' dynamics. You first looked at how FSIG turbines fit into a network scheme regarding network damping. Then you reviewed how DFIG turbines integrate into a grid. You read about the similarities and differences between these generator options. Specifically, you studied their impact on grid power and voltage control. The next section covered some specific issues with connecting wind turbines to large networks. You read about issues like frequency, active power control, short-circuit voltage, reactive power control, voltage flicker, harmonics, and network stability. You also read about how utilities and regulators deal with problems in these areas of grid integration. The chapter ended with some information about substations. You read about the main types of substations and their functions. You also reviewed the equipment that each kind of substation has to complete its functions.

CHAPTER KEY CONCEPTS AND TERMS

Bus bar

Collector substation

Distribution substation

Electrical substation

Flickermeter

Mega volt-amperes

Periodical functions

Total harmonic distortion (THD)

Transmission substation

Voltage flicker

CHAPTER ASSESSMENT: WIND FARMS

1. Which of the following descriptions about FSIG and DFIG turbines' influence on grid damping is the most precise?
   - ❏ **A.** DFIG turbines add to network damping via torque synchronization. FSIGs affect network damping via use of torque damping.
   - ❏ **B.** DFIG turbines affect network damping via bus bar voltage regulation. FSIGs use torque synchronization and torque damping.
   - ❏ **C.** DFIG and FSIG turbines affect network damping via torque synchronization and damping, but they use different control schemes.
   - ❏ **D.** DFIG and FSIG turbines both affect network damping via the FMAC control method.

2. In simple operational network damping models, it is safe to assume that bus bar voltage levels are _____.

3. DFIG _____ _____ have much less influence on power variations than do those in FSIGs.

4. In DFIGs, _____ _____ responds very quickly to rotor voltage changes.

5. Nowadays, most wind system grid codes are fulfilled by using which of the following?
   - ❑ **A.** Magnetization control methods
   - ❑ **B.** Power electronic converter methods
   - ❑ **C.** Active and reactive power control methods
   - ❑ **D.** None of the above

6. What is the source of wasted energy when considering network frequency and active power?
   - ❑ **A.** The source of waste is synchronous generators running at suboptimal levels to leave room for frequency increases if network voltage drops.
   - ❑ **B.** The source of waste is asynchronous generators running at suboptimal levels to leave room for frequency increases if network voltage drops.
   - ❑ **C.** The source of waste is synchronous generators producing excess power, which has to be absorbed back into the generators.
   - ❑ **D.** The source of waste is asynchronous generators producing excess power, which must be stored for later use if network voltage drops.

7. Short-circuit levels are directly related to voltage quality, and they provide a good measure of network strength.
   - ❑ **A.** True
   - ❑ **B.** False

8. When considering short-circuit situations in a network, the variable Sk stands for which of the following?
   - ❑ **A.** Short-circuit tolerance in mega volt-amperes
   - ❑ **B.** Short-circuit tolerance in megawatts
   - ❑ **C.** Impedance between two given points in a grid
   - ❑ **D.** Nominal voltage in a remote part of the grid from the shorted circuit

9. A grid with small impedance is considered to be a strong grid.
   - ❑ **A.** True
   - ❑ **B.** False

10. Changes to which of the following items in induction generators allow them to produce and consume reactive power?
    - ❑ **A.** PWM converter settings
    - ❑ **B.** Energy oscillation levels
    - ❑ **C.** Magnetization levels
    - ❑ **D.** Rotor phase angle

11. Which of the following items is used for most harmonics control in turbine generators?
    - ❑ **A.** PWM converters
    - ❑ **B.** Power electronic converters
    - ❑ **C.** Harmonic filters
    - ❑ **D.** Pulse-width switching converters

12. If there is a loss of generation capacity, low- to no-voltage situations will occur unless which of the following occurs?
    - ❏ **A.** Remaining operative generators are running at full capacity.
    - ❏ **B.** Network impedance levels were very low before the generation loss.
    - ❏ **C.** Remaining operative generators are running at reduced capacity.
    - ❏ **D.** The network was in an overload situation when the generation loss happened.

13. Why do many utilities require that wind farms contain ride-through control systems?
    - ❏ **A.** They help prevent power losses during short-circuit situations.
    - ❏ **B.** They avoid making short-circuit situations in a grid worse.
    - ❏ **C.** They help minimize distribution interruptions in overload situations.
    - ❏ **D.** They contribute to lower impedance levels in a power grid.

14. Transmission substations are primarily responsible for stepping voltage up or down as required by grid load.
    - ❏ **A.** True
    - ❏ **B.** False

15. The main difference in equipment between collector substations and distribution substations is that collector substations often use _____ _____.

chapter 10

Power System Stabilizers and Wind Farm Network Damping

THIS CHAPTER IS ABOUT power system stabilizers (PSSs). These are generator controllers that help with two key areas. The first is system damping in oscillatory conditions. The second is network transience after a fault. In this chapter, you will read about power system stabilizers in general. Then you will compare how power system stabilizers are used on three different kinds of generators. First, you will read how they are used with synchronous generators. This includes their effects on system damping and network transience. Then, you will study the same issues with doubly fed induction generators (DFIGs). Lastly, you will review power system stabilizer use on fully rated converter-based generators (FRC generators). When you finish this chapter, you should be able to discuss the basic differences between power system stabilizers on all three of these generator types.

Chapter Topics

This chapter covers the following topics and concepts:

- Power system stabilizer overview
- Synchronous generator PSSs
- Doubly fed induction generator PSSs
- Fully rated converter-based generator PSSs

Chapter Goals

When you complete this chapter, you will be able to:

- Describe the main uses of power system stabilizers on synchronous, doubly fed induction, and FRC generators
- Explain the characteristics of synchronous generator PSS use
- Dialogue on synchronous generator PSS influence on network damping and network transient performance
- Explain the characteristics of DFIG PSS use
- Describe the DFIG PSS influence on network damping and network transient performance
- Explain the characteristics of FRC generator PSS use
- Dialogue on FRC generator PSS influence on network damping and network transient performance
- Compare and contrast the various benefits of PSS use on synchronous, doubly fed induction, and FRC generators on network transient performance and system oscillatory damping

Power System Stabilizer Overview

Power system stabilizers are also called PSSs. They are electronic control circuits for oscillatory problems due to negative damping. In the United States, high generator outputs are transmitted over long distances. This contributes to negative damping conditions, a condition that is unstable and causes the voltage to increase in an undesirable fashion. PSSs modulate a voltage regulator's reference inputs via an additional signal. This produces positive damping torque, thereby maintaining the voltage at a desired level. Different generator types use PSSs in different ways. Read on to learn more about PSS use in three different generator types.

NOTE

Power system stabilizers are an important part of the power grid connection for wind turbines. Uncontrolled voltages can lead to equipment damage in the grid, in the wind turbine or even in consumer equipment. Maintaining system-wide voltages and clean power is very important to optimal performance of all components on and connected to the power grid.

Synchronous Generator Power System Stabilizers

Synchronous generator PSSs improve damping by changing field voltage. On the basis of system oscillations, these generators' power variations synchronize to rotor system oscillations. The PSS input signal can be any variable responding to system oscillations. PSS output is added to the automatic voltage

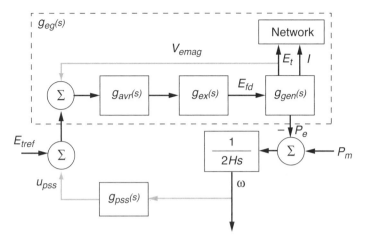

FIGURE 10-1 Synchronous generator PSS excitation control scheme.

Adapted from Anaya-Lara, O., Jenkins, N., Ekanayake, J., Cartwright, P., & Hughes, M. (2009). *Wind energy generation: Modelling and control.* West Sussex, England: John Wiley & Sons, Ltd.

regulator (AVR) excitation control loop's reference set point. The most common input signals are rotor speed and generator electrical power **FIGURE 10-1**.

Generator response based on field and power variations changes due to two factors. The first is operating conditions. The second is system load. Network oscillation frequency changes with generator and network operations. Frequency band damping action covers local and inter-area oscillations. PSS phase compensation must add positively to damping. This is true across the entire frequency bandwidth and operational range.

You can deduce that if phase lead is below optimum, a PSS adds negatively to synchronizing power. So if phase lead is above optimum, a PSS adds positively to synchronizing power.

You can consider governor actions causing power variations to be stable when studying PSS models. This is because governor actions are slow enough to assume this. In oscillatory situations, electrical power lags speed by 90 degrees. If electrical power is the PSS input signal, a negative gain results. Then either **phase lag compensation**, or simply adjustment of electrical power relative to the speed of the turbine rotation, is needed. Phase lag compensation is a control system component that helps eliminate undesirable frequency responses in feedback and control systems. As frequency oscillations fall, the phase lag falls. So the compensator should increase phase lag to keep combined compensator lag stable. This is at about 90 degrees.

In synchronous generators, AVRs and PSSs both function based on changing the same variable. The variable is excitation control. Thus, independent damping and voltage control is impossible. PSSs can better damping and extend operating regions, but doing this comes at the expense of voltage control. Thus, voltage recovery is slower after system faults.

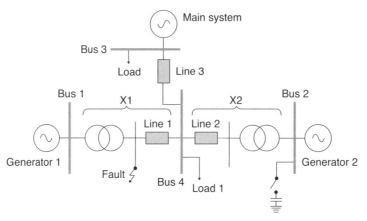

FIGURE 10-2 Simple synchronous generator damping model.

Adapted from Anaya-Lara, O., Jenkins, N., Ekanayake, J., Cartwright, P., & Hughes, M. (2009). *Wind energy generation: Modelling and control.* West Sussex, England: John Wiley & Sons, Ltd.

Synchronous Generator PSS Influence on Damping

PSS damping and transience influence is based on the simple model shown in FIGURE 10-2 . FIGURE 10-3 shows that PSS use improves network damping. This PSS use is based on electrical power signal input. In Figure 10-3, Generator 1 has a capacity of 2,800 MVA. Operations for Generator 2 show operating step increases from 0 to 2,160 MVA. This is shown in 20 percent intervals. Generator 2's total capacity is 2,400 MVA. Disregard Generator 1's PSS. Increasing Generator 2's

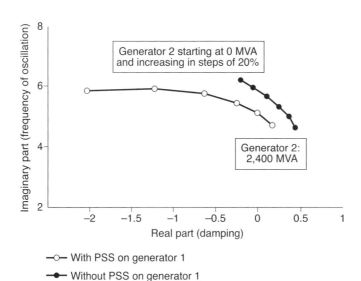

FIGURE 10-3 Model of Generator 1 PSS network damping influence.

Data from Anaya-Lara, O., Jenkins, N., Ekanayake, J., Cartwright, P., & Hughes, M. (2009). *Wind energy generation: Modelling and control.* West Sussex, England: John Wiley & Sons, Ltd.

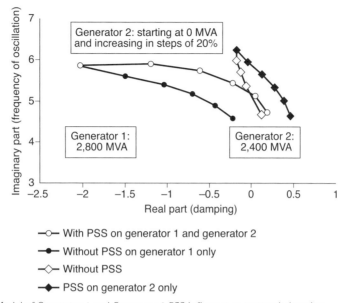

FIGURE 10-4 Model of Generator 1 and Generator 2 PSS influence on network damping.

Data from Anaya-Lara, O., Jenkins, N., Ekanayake, J., Cartwright, P., & Hughes, M. (2009). *Wind energy generation: Modelling and control.* West Sussex, England: John Wiley & Sons, Ltd.

capacity has a large effect on system damping. Generator 1's PSS greatly improves network damping if Generator 2's output is low. As Generator 2's capacity rises, PSS damping becomes less pronounced. This leaves the system unstable.

In **FIGURE 10-4** both generators use PSSs, network dynamic stability is much improved. This allows both generators to run at near-maximum capacity under stable conditions.

Synchronous Generator PSS Influence on Transient Operations

See Figure 10-2 again. Assume the same network model for studying PSS influence on transient operations. Consider an 80 ms three-phase fault applied on line 1. It is close to Generator 1's transformer terminal.

Figure 10-3 shows a low-capacity (480 MVA) case with Generator 2 having no PSS. The network is stable but has low damping. **FIGURE 10-5** shows the transient responses after the fault. Both generators desynchronize from the grid. **Pole slipping** also occurs. Pole slipping occurs when a rotor field shifts its position relative to the stator field. This can be a serious problem. It can cause generator damage. In extreme cases, it can destroy a generator.

Now assume Generator 1 contains a PSS. After the fault, both generators stay synchronized with the network. They achieve fault ride-through. Now assume Generator 2's capacity rises to 960 MVA. Generator 1 uses a PSS. Here, the

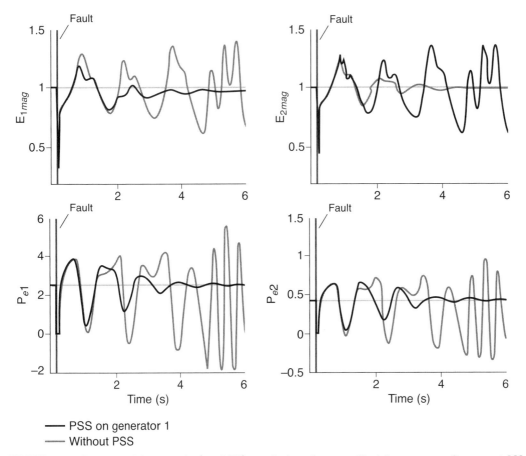

FIGURE 10-5 Generator 2's low capacity (480 MVA) post-fault performance. Black lines represent Generator 1 PSS use; gray lines represent Generator 1 without PSS use.

Data from Anaya-Lara, O., Jenkins, N., Ekanayake, J., Cartwright, P., & Hughes, M. (2009). *Wind energy generation: Modelling and control.* West Sussex, England: John Wiley & Sons, Ltd.

generators desynchronize from the network. If Generator 2 uses a PSS in the same situation, both achieve fault ride-through **FIGURE 10-6** .

PSS use on both generators improves transient operations and network damping, but if Generator 2's capacity rises to 1,440 MVA and both generators have PSSs, desynchronization occurs after fault clearance. This model indicates using only synchronous generators places high demand on excitation control.

Doubly Fed Induction Generator Power System Stabilizers

DFIG network damping increases by adding an auxiliary PSS loop. In oscillatory conditions, PSSs inject power variations into the grid. These spur extra damping in synchronous generators. In theory, the PSS input signal can be any DFIG variable responding to grid oscillations. This includes rotor speed, slip, or stator

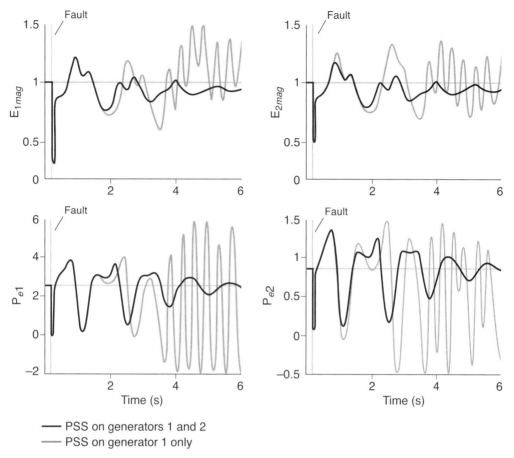

FIGURE 10-6 Generator 2 at 960 MVA capacity post-fault performance. Black lines represent Generator 1 PSS use; gray lines represent generators 1 and 2 with PSS use.

Data from Anaya-Lara, O., Jenkins, N., Ekanayake, J., Cartwright, P., & Hughes, M. (2009). *Wind energy generation: Modelling and control.* West Sussex, England: John Wiley & Sons, Ltd.

electrical power. u_{pss} (PSS output signal) is added to the power control loop's reference set point. **FIGURE 10-7** shows the flux magnitude and angle controller (FMAC) control scheme. As in synchronous generators, DFIG PSSs consist of a **washout term**. A washout term is the upper limit on the gain that the differentiator output can have as the frequency of the input signal increases in a control system. It eliminates steady-state offset. It also consists of a phase shift compensator. This aligns PSS output with the necessary phase relationship. This is key to network damping reduction.

DFIG contributions to network damping require power oscillation injection. These stimulate power variations in synchronous generators in phase with their rotor speed oscillations. DFIGs using FMAC control schemes create **lag term**, which is an offsetting adjustment to the phase of the generators' excitation. It is

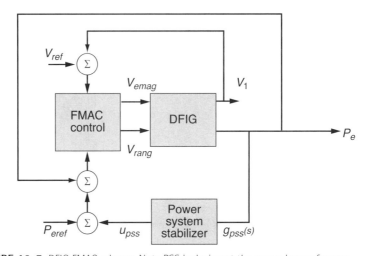

FIGURE 10-7 DFIG FMAC scheme. Note PSS inclusion at the power loop reference.

Adapted from Anaya-Lara, O., Jenkins, N., Ekanayake, J., Cartwright, P., & Hughes, M. (2009). *Wind energy generation: Modelling and control.* West Sussex, England: John Wiley & Sons, Ltd.

located in the power loop, as shown in **FIGURE 10-8**. The lag term ensures small, positive network damping. If PSS control augments the power vector and rotates counterclockwise, DFIG damping increases.

DFIG PSS Influence on Damping

To study DFIG PSS damping, think of the previous model. Generator 2 is a DFIG with FMAC control. The PSS input signal is stator power. The PSS output is applied

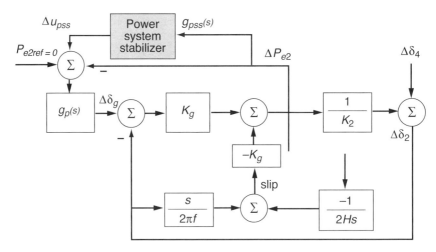

FIGURE 10-8 Reduced block diagram of DFIG with stator-driven PSS.

Adapted from Anaya-Lara, O., Jenkins, N., Ekanayake, J., Cartwright, P., & Hughes, M. (2009). *Wind energy generation: Modelling and control.* West Sussex, England: John Wiley & Sons, Ltd.

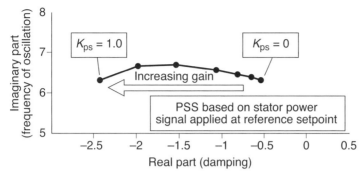

FIGURE 10-9 Influence of PSS gain (DFIG with FMAC control, 0 to 2,400 MVA) on Kps.

Data from Anaya-Lara, O., Jenkins, N., Ekanayake, J., Cartwright, P., & Hughes, M. (2009). *Wind energy generation: Modelling and control.* West Sussex, England: John Wiley & Sons, Ltd.

at the DFIG power control loop's reference set point. **FIGURE 10-9**. The plots show the PSS influence increase from a capacity of 0 to 2,400 MVA. At 2,400 MVA, the DFIG operates at a slip value of -0.1. As the **gain** increases, the system oscillation values shift steadily left up to a value of $K_{ps} = 1.0$. Gain is the increase (or decrease if negative), as a multiplicative factor of some variable, such as voltage, current, or frequency. Further increase is possible with a continuing shift to the left, but these values are shown for operating situations where other network and generator values stay near optimal.

TECH TIPS

Having the right test equipment is important for technicians in order to diagnose problems in PSS and connected systems. Check with the turbine manufacturer to ensure you have the right electronic test equipment to measure dampening and transient problems that may be encountered.

   FIGURE 10-10 shows the PSS damping influence as generator capacity increases. The greater the DFIG's capacity, the more influence its PSS has on system damping in synchronous generators. **FIGURE 10-11** shows how DFIG slip values influence PSS damping via power contributions. As DFIG operating speed decreases, slip values go from supersynchronous to subsynchronous. So system damping also decreases. DFIGs use higher wind speeds to increase

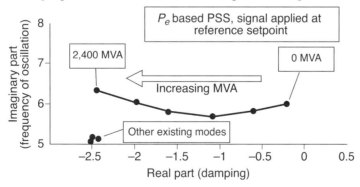

FIGURE 10-10 DFIG with FMAC control; Generator 2 PSS influence on network damping.

Data from Anaya-Lara, O., Jenkins, N., Ekanayake, J., Cartwright, P., & Hughes, M. (2009). *Wind energy generation: Modelling and control.* West Sussex, England: John Wiley & Sons, Ltd.

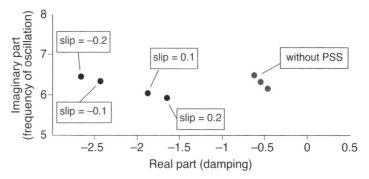

FIGURE 10-11 DFIG with FMAC control; generator slip influence on network damping.

Data from Anaya-Lara, O., Jenkins, N., Ekanayake, J., Cartwright, P., & Hughes, M. (2009). *Wind energy generation: Modelling and control.* West Sussex, England: John Wiley & Sons, Ltd.

operating speeds. This improves their power transfer efficiency. At lower wind speeds and operating speeds, DFIG system damping decreases.

DFIG PSS Influence on Transient Operations

Consider the same DFIG using FMAC control and a PSS. Application of an 80 ms three-phase fault helps study PSS influence on network transience. The fault occurs on line 1 near Generator 1's transformer terminals.

In **FIGURE 10-12**, the dotted lines show DFIG operations with no PSS control. The operating slip value is –0.1. The DFIG sustains fault ride-through, but system oscillations continue more than 5 seconds after fault clearance. Compare that with the full lines. They represent the same operating conditions using PSS control. The generator achieves fault ride-through. After fault clearance, system damping is normalized quickly. The DFIG increases power variations, but there is little change in voltage recovery. The synchronous generator without PSS control did not achieve fault ride-through. This is a key difference between synchronous generators and DFIGs.

FIGURE 10-13 shows the same post-fault operation, but the operating slip value is 0.1. The post-fault damping improves with PSS use, but in this case, operating power level is much lower. Immediately after fault clearance, there are high signal levels. PSS controller limits impinge its output magnitude. Over the recovery period, the post-fault DFIG performance, no matter the slip values, is similar.

Fully Rated Converter–Based Generator Power System Stabilizers

FRC wind farms use limited, fast control of voltage and power. This does not allow for system damping. In normal operations, terminal voltage magnitude is constant. Power output varies based on wind speeds. The goal of power control is

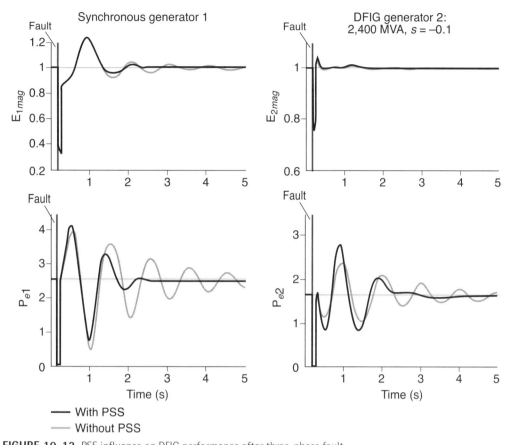

FIGURE 10-12 PSS influence on DFIG performance after three-phase fault.

Data from Anaya-Lara, O., Jenkins, N., Ekanayake, J., Cartwright, P., & Hughes, M. (2009). *Wind energy generation: Modelling and control.* West Sussex, England: John Wiley & Sons, Ltd.

keeping system operations stable. It provides no network damping. If an FRC wind farm must contribute to system damping, it uses PSSs. Consider an FRC operating at steady-state levels. It has constant generator output power and voltage levels. Bus bar voltages are assumed constant. So line voltages are based only on bus bar voltage phase changes.

The goal of FRC control is keeping output constant. Local variables like voltage magnitude, current magnitude, and power output do not influence damping. In an FRC, PSS control is based on network frequency variations. This is the input signal. This signal is added to the grid-side converter's output power reference set point. It is also added to that of the generator-side converter.

FRC PSS Influence on Damping

In **FIGURE 10-14**, Generator 1 operates synchronously at 2,800 MVA. Generator 2 is an FRC. It runs at 2,400 MVA. Without PSS control, the plots lie on the right

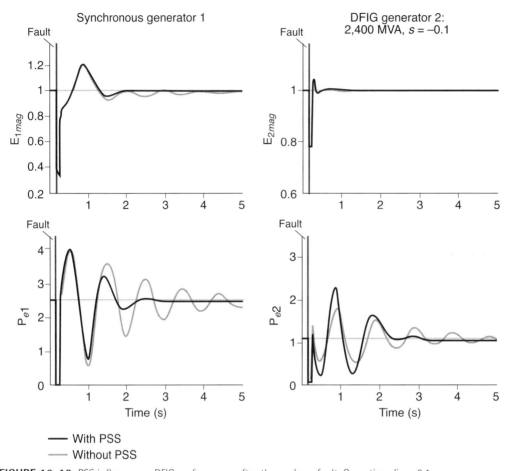

FIGURE 10-13 PSS influence on DFIG performance after three-phase fault. Operating slip = 0.1.

Data from Anaya-Lara, O., Jenkins, N., Ekanayake, J., Cartwright, P., & Hughes, M. (2009). *Wind energy generation: Modelling and control.* West Sussex, England: John Wiley & Sons, Ltd.

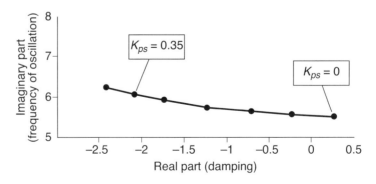

FIGURE 10-14 FRC PSS gain influence on local mode plots.

Data from Anaya-Lara, O., Jenkins, N., Ekanayake, J., Cartwright, P., & Hughes, M. (2009). *Wind energy generation: Modelling and control.* West Sussex, England: John Wiley & Sons, Ltd.

side of the grid. This indicates network dynamic instability. As the FRC PSS gain increases, the plots move left. This shows that the PSS has a positive influence on synchronous Generator 1's damping.

PSS design for FRCs must regard large disturbances after network faults. It also must provide damping contributions to the grid. Higher PSS gains lead to more impact on converters' DC voltage levels. If higher gain and higher damping are needed, extra controls keep DC voltage variation within normal limits. At any installed FRC capacity, FRC PSS damping increases with generation power level increases.

FRC PSS Influence on Transient Operations

The same network model shows FRC PSS influence on network transience. The same fault on line 1 is also used. The PSS qualities discussed in the damping section are also the same.

FIGURE 10-15 assumes Generator 1 is synchronous. It has 2,800 MVA of generation capacity. It uses only AVR control. Generator 2 has 2,400 MVA of FRC

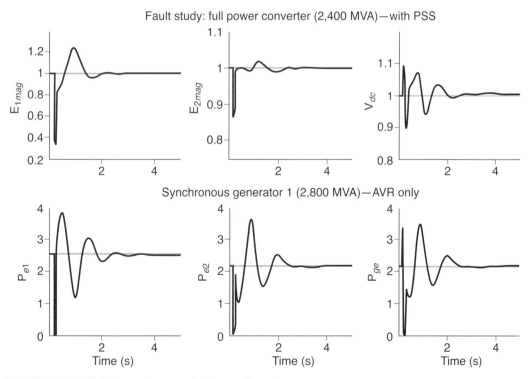

FIGURE 10-15 FRC PSS control on post-fault network transience.

Data from Anaya-Lara, O., Jenkins, N., Ekanayake, J., Cartwright, P., & Hughes, M. (2009). *Wind energy generation: Modelling and control.* West Sussex, England: John Wiley & Sons, Ltd.

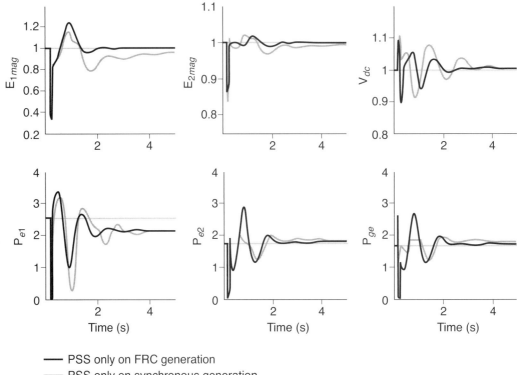

— PSS only on FRC generation
— PSS only on synchronous generation

FIGURE 10-16 FRC PSS performance on post-fault network transience.

Data from Anaya-Lara, O., Jenkins, N., Ekanayake, J., Cartwright, P., & Hughes, M. (2009). *Wind energy generation: Modelling and control.* West Sussex, England: John Wiley & Sons, Ltd.

capacity with a PSS. This combination achieves fault ride-through. It also provides good synchronous generator damping. Compare this with an FRC without PSS control. Without PSS control, FRCs can only achieve fault ride-through at a maximum of 960 MVA.

FIGURE 10-16 compares PSS additions with network transience. The dotted lines show only FRC PSS use. The full lines show only synchronous PSS use. The synchronous generator has 2,400 MVA of generation capacity. The FRC generator has 1,960 MVA of capacity. In both cases, the synchronous generator also uses AVR control. The FRC with PSS control provides better transient performance for both generators after fault clearance. Generator 1 provides better oscillation damping and voltage recovery directly after fault clearance. The FRC PSS generation has smaller terminal voltage magnitude. It also has fewer DC link voltage variations.

INTERCONNECTION OF RENEWABLE ENERGY SYSTEMS

The United States is gaining more and more benefits from wind energy. Grid operators must understand how to integrate wind power into traditional systems. The US Department of Energy's Wind and Water Power Program works with grid operators, utilities, and regulators. They are creating strategies for adding wind energy into power systems while keeping them stable. Many utilities have concerns about wind energy's impacts on their existing systems.

The program's goal for interconnecting renewable systems is to address electric power market rules, connection impacts, operating strategies, and needed system planning. It hopes to complete this by the end of 2012. Some of the main program areas include the following:

- **Research projects**—Program researchers work with industry players. Their research projects increase utilities' understanding of integration issues. They also raise confidence in new wind energy products.
- **Wind forecasting**—Improving current forecasts is a beneficial strategy for eliminating uncertainty. The program provides forecast verification data. This helps improve forecast accuracy. The program provides accurate wind data for modern turbine heights.
- **Wind plant performance characterization**—The program also collects data about power supplied by current wind farms. These data help determine wind farm power curves. They also improve wind forecasts and turbine wake evaluation.
- **Grid planning**—The program informs system planners on efficiently representing wind power's characteristics. The program sponsors large-scale integration studies. They analyze transmission needs. They also examine operational and production cost impacts of renewable energy source usage.
- **Grid operational impact analysis**—Without realistic analysis and cost allocation, utilities overestimate operational costs. This results in undervaluing of wind power. Program research uses engineering and cost analyses. These quantify and allocate cost impacts. Short- and long-term areas of concern include plant forecasting and control, energy storage, and regional cooperation. These studies help reduce integration costs.
- **Outreach and education**—Outreach and education inform many parties about wind energy integration. The parties include corporate utilities, cooperatives, public power companies, regulatory groups, and standard-setting groups. Program outreach circulates research analyses, results, and best practices. These efforts encourage wind power inclusion in system portfolios. It also ensures continued wind energy expansion.
- **Integration of wind energy and hydropower technologies**—The synergy between wind and hydropower integration is not yet fully realized. Researchers work with federal Power Marketing Administrations. They analyze potential and existing hybrid projects. They also analyze watershed basin and electric control areas. This cooperation determines if hydropower can combine with wind energy without affecting other flow requirements. It also clarifies the benefits of hybrid system integration.

CHAPTER SUMMARY

This chapter was about power system stabilizers. They are generator controllers that help with two key areas. The first is system damping in oscillatory conditions. The second is network transience after a fault. You began by learning a little about power system stabilizers in general. Then you moved on, comparing how power system stabilizers are used on three different kinds of generators. First you read how they are used with synchronous generators. This includes their effects on system damping and network transience. Then you studied the same issues with DFIGs. Lastly, you reviewed power system stabilizer usage on FRC generators. You should now be able to discuss the basic differences between power system stabilizer use on these three generator types.

CHAPTER KEY CONCEPTS AND TERMS

Gain
Lag term
Phase lag compensation
Pole slipping
Power system stabilizers (PSSs)
Washout term

CHAPTER ASSESSMENT: POWER SYSTEM STABILIZERS AND WIND FARM NETWORK DAMPING

1. Power system stabilizers help with system damping of oscillations caused by which of the following?
   ❏ A. Transmission of power over short distances
   ❏ B. Transmission of power between grids
   ❏ C. Transmission of power over long distances
   ❏ D. Transmission of power through electric substations

2. Synchronous generator PSSs improve _____ _____ by changing field voltage.

3. In a synchronous generator, if phase lead is below optimum, what effect does the PSS have on synchronizing power?
   ❏ A. It adds negatively to synchronizing power.
   ❏ B. It adds positively to synchronizing power.
   ❏ C. It has no effect on synchronizing power.
   ❏ D. None of the above

4. In the synchronous PSS model for studying network damping, Generator 1's PSS greatly improves network damping if Generator 2's output is _____.

5. Which of the following is the best definition for the term *pole slipping*?
   - ❑ **A.** It occurs when the stator field shifts its position relative to the rotor field.
   - ❑ **B.** It occurs when the rotor field shifts its position relative to the permanent magnet.
   - ❑ **C.** It occurs when the rotor field shifts its position relative to the stator field.
   - ❑ **D.** It occurs when the permanent magnet shifts its position relative to the rotor field.

6. What is the main effect that adding a PSS to a DFIG achieves?
   - ❑ **A.** It amplifies the steady-state offset.
   - ❑ **B.** It stimulates network damping from synchronous generators connected to the grid.
   - ❑ **C.** It contributes to DFIG efforts to influence system damping.
   - ❑ **D.** Both B and C

7. As DFIG operating speed increases, slip values go from supersynchronous to subsynchronous.
   - ❑ **A.** True
   - ❑ **B.** False

8. PSS controller limits impinge its output magnitude. Over the recovery period, the post-fault DFIG performance, no matter the _____ _____, is similar.

9. The overall aim of FRC generator control is to keep output _____.

10. In FRC generators, if higher gain and higher damping are needed, extra controls keep _____ _____ _____ within normal limits.

11. Which of the following items is used for most harmonics control in turbine generators?
    - ❑ **A.** PWM converters
    - ❑ **B.** Power electronic converters
    - ❑ **C.** Harmonic filters
    - ❑ **D.** Pulse-width switching converters

12. In the FRC network model you studied, without PSS control FRCs achieve fault ride-through only at a maximum capacity of _____.

Answer Key

CHAPTER 1   Wind Energy Generation and Conversion

1. D 2. A 3. C 4. A and B 5. A 6. Vertical axis turbines
7. A 8. D 9. D 10. A 11. D 12. B and D 13. C

CHAPTER 2   Modern Power Electronics and Converter Systems

1. B 2. One to 48 hours 3. D 4. B 5. A 6. C 7. A 8. A
9. B 10. B 11. Diode bridge circuits 12. DC to AC
13. D 14. Self-commutated converters 15. Pure sine wave

CHAPTER 3   Fixed–Speed Induction Generators

1. B 2. Asynchronous generators 3. C 4. B
5. Elevated voltage 6. A 7. Insulated 8. B 9. Windings 10. B

CHAPTER 4   Synchronous Generators for Wind Turbines

1. B 2. B 3. Two 4. Concentrated, nonuniform 5. Two-thirds
6. 120-degree 7. Synchronous 8. A 9. Salient-pole 10. B
11. A 12. A 13. B 14. Resistance, reactance 15. Droop

CHAPTER 5   Doubly Fed Induction Generators

1. A 2. Wound-rotor 3. Variable-speed
4. Generator stator, voltage source converters 5. B 6. A
7. B 8. B 9. Copper 10. Power converters 11. B
12. Electromagnetic torque 13. A 14. B 15. B

CHAPTER 6   Fully Rated Converter–Based Generators

1. B 2. D 3. Power loss 4. A 5. B 6. B 7. Higher 8. D 9. B
10. DC link voltage 11. D 12. A 13. A 14. Vector 15. d

CHAPTER 7   Wind Turbine Control

1. B 2. 1 percent 3. C 4. 3, 5 5. B 6. B 7. Reduced 8. B
9. A 10. DFIG 11. B 12. B 13. B 14. HVDC 15. Gearbox

CHAPTER 8   Rotor Dynamics

1. C 2. In-plane 3. A 4. Total hub inertia 5. A 6. C
7. A 8. B 9. B 10. C 11. A 12. C 13. A
14. Two-mass model 15. 5, 200

CHAPTER 9   Wind Farms

1. C 2. Constant 3. Slip variations 4. Transient reactance
5. B 6. A 7. B 8. A 9. A 10. C 11. C 12. C 13. B
14. B 15. HVDC converters

CHAPTER 10   Power System Stabilizers and Wind Farm Network Damping

1. C 2. Network damping 3. A 4. Low 5. C 6. D
7. B 8. Slip values 9. Constant 10. DC voltage variation
11. C 12. 960 MVA

Active power   The amount of power generated for load consumption.

Anemometer   Outside the nacelle, a small turbine that measures wind speed and transmits wind speed data to the controller.

Armature winding   A copper winding located on the stator. This is a three-phase winding and creates a magnetic field that interacts with that of the rotor winding.

Automatic voltage regulators (AVRs)   These keep the stator's terminal voltage within defined limits. When increased demand causes terminal voltage to fall, the AVR augments terminal voltage.

Betz limit   A law that states that no wind turbine can convert more than 16/27 (59.3 percent) of the kinetic energy of wind into useful energy.

Bus bar   A heavy-duty conducting rod that carries electrical voltages between loads and supply systems. Bus bars often form part of transformer station equipment for electricity distribution.

Collector substation   A substation that is similar to a distribution substation, but is usually located within a wind energy production system. Such a substation collects power generated by individual turbines and steps it up to the voltage level required for high-voltage transmission into the grid.

Contactors   Relays that switch large currents into their system. They usually contain many switches and most are the "normally open" type.

Controller   Housed in the nacelle, the controller starts the turbine at 8–16 mph and shuts down at 55 mph to prevent high wind damage.

Dead band   A voltage range in a generator's governor where no power output changes are possible. Generators produce more power when their voltages fall outside this voltage range.

Disk brake   Housed in the nacelle, the disk brake (mechanical, electrical, or hydraulic) stops the rotor in emergencies.

Distribution substation   Substation that exists to reduce voltage levels before delivery to the majority of consumers. Unless a consumer uses large amounts of high-voltage electricity, it is not economical or safe to connect directly to the high-voltage transmission system.

Drive shaft   The main cylinder connecting a wind turbine's rotor to its electricity-generating component. The drive shaft carries mechanical energy extracted from the wind, usually through a gearbox, and delivers it for transformation to electric energy.

Droop governors   Electrical components that often work to control generator frequency. They work by decreasing generator equilibrium speed when loads increase.

Electrical substation   Network location where voltage levels are converted from higher to lower or from lower to higher. They also provide switching, control, and other functions.

Electric crowbar   An electric circuit that intentionally shorts out in case of power surges. Electric crowbars protect the electrical components of the generators.

Electric torque   The power balance in the gap between stator and rotor. This is derived from the following formula: $T_e = I_r^2 \left(\dfrac{r_r}{s} \right) + \dfrac{P_r}{s}$.

Electromechanical relays   Devices that both complete and interrupt circuits. They do so by making physical contact between two points that are electrically charged.

Excitation   The interaction between static and rotating magnetic poles.

Excitation control system   A system that provides automatic voltage regulation and protects the generator from excess power surges.

Exciters   Electric system components that supply adjustable direct currents to a generator's field winding. These can even be generators in their own right, especially in small-scale operations.

Field winding   A copper-based winding located on the rotor. It creates a magnetic field that interacts with that of the armature winding.

Flickermeter   A device that measures light waves in the spectrum visible to the human eye. They are useful for establishing P_{st} and P_{lt} values in a network.

Gain   The increase (or decrease if negative), as a multiplicative factor, of some variable such as voltage, current, or frequency.

Gearbox   Housed in the nacelle, the gearbox connects the low-speed shaft to the high-speed shaft and increases the rotational speeds from 30–60 rpm to 1,000–1,800 rpm to produce electricity.

Generator   The generator is usually of the off-the-shelf induction type and produces 60-cycle AC electricity. The high-speed shaft drives the generator. The generator is contained in the nacelle (atop the tower).

Harmonic distortion   The result of waveform changes caused principally by inverters, motor drives, electronic appliances, light dimmers, fluorescent light ballasts, and personal computers.

Harmonic filters   Filters that are necessary to minimize the effects of harmonic distortion. These work by improving the waveform closer to a pure sine wave and they reduce harmonics' side effects.

Horizontal axis wind turbine (HAWT)   A two- to three-blade wind turbine that operates upwind (toward wind) and contains a gearbox. The main rotor shaft and electrical generator are located at the top of the tower and face upwind.

Inverters   Components that convert DC electrical loads to AC electrical loads. Electronic inverters are made of circuit elements and control circuitry. The circuit elements switch high currents and the circuitry coordinates the switching.

In-plane bending   Blade motion within the rotor plane.

Kinetic energy   Kinetic Energy = 0.5 × Mass × Velocity^2, where kinetic energy is measured in joules, mass is measured in kilograms (1 kg = 2.2 lbs.), and the velocity is measured in meters per second (1 meter = 3.281 ft. = 39.37 inches).

Kirchhoff's circuit law   A law that states that the total current entering a node is equal to the current leaving the node.

Kirchhoff's voltage law   A law that states that the sum of voltages around a loop equals zero.

Lag term   An offsetting adjustment to the phase of the generators' excitation.

Limiter switches   A switch that is most similar to a light switch, but that can close or open circuits. The main function is via a visible device connected to an electromagnetic actuator.

Load compensators   Additional protective loops built in to AVRs. This loop allows grid voltage to be controlled from a remote point.

Low-voltage protection (LVP)   Refers to controllers that depower motors in low-voltage conditions. They also keep the motors from restarting when normal voltage returns.

Low-voltage release (LVR)   Refers to controllers that also depower motors in low-voltage situations.

Low-voltage release effect (LVRE)   Refers to a type of manual controller that holds motors constantly at full voltage.

Mega volt-amperes   The measurement unit used for electrical transformer capacity.

Nacelle   The nacelle houses the generator, gearbox, low- and high-speed shafts, controller, and brake.

Nearshore   Nearshore turbines are usually located closer to the shore (within 3 km) or on water within 10 km of land; convection increases wind speeds as the heating of land and sea differentiate.

Negative sequence   The energy flux resulting from currents flowing opposite to the rotor's rotational direction.

Normally closed, timed closed (NCTC) relays   Also called "normally closed, off-delay" relays. This type of relay is also closed when the coil is de-energized. As soon as the coil is energized, the circuit opens. However, when the coil is depowered, the reclosing of the circuit is delayed.

Normally closed, timed open (NCTO) relays   Also called "normally closed, on-delay" relays. This type of relay is closed when the coil is de-energized. When the coil is powered for a specific amount of time, the armature opens the circuit. As soon as the coil loses power, the circuit recloses.

Normally open, timed closed (NOTC) relays   Also called "normally open, on-delay" relays. Usually this type of relay is open when the coil is de-energized. When the coil has been powered for a specific amount of time, the armature closes the circuit.

Normally open, timed open (NOTO) relays   Also called "normally open, off-delay" relays. This type of relay is also open when the coil is de-energized. However, when the coil is energized in this type of relay, the armature closes the circuit immediately.

Occasional service   Power delivered no more than 10 seconds after a frequency drop. It is expected to last from 20 seconds to 30 minutes after initial response time.

Offshore   Refers to turbines that are at least 10 km away from land. The distance reduces noise pollution and aesthetic environment concerns; however, utilization rates, transportation, installation, and maintenance costs are much higher. Offshore wind farm technology involves either a fixed-bottom, foundation-based tower or a deep-water, floating turbine.

Onshore   Onshore turbines tend to be in hilly, mountainous, or ridgeline-like areas, about 3+ km from the nearest shore; wind accelerates over ridges.

Out-of-plane bending   Blade bending perpendicular to the rotor plane. It is normal within the direction of rotor rotation.

Periodical functions   Functions that repeat over and over and can be expressed in mathematical terms.

Phase lag compensation   A control system component that helps eliminate undesirable frequency responses in feedback and control systems.

Pitch angle power control   A method of control that depends on standard power control for feathering turbine blades. In this control method, the blade pitch is gradually reduced as wind speed increases.

Pole pitch   The distance between poles based on angular separation. Pitch reduction helps to maximize generator efficiency.

Pole slipping   This occurs when a rotor field shifts its position relative to the stator field. It can be a serious problem, causing generator damage or destruction.

Power converters   Converters that convert electricity from one form to another.

Positive sequence   The energy flux resulting from currents flowing in the direction of rotor rotation.

Power system stabilizers (PSSs)   The electronic control circuits for oscillatory problems due to negative damping.

Prime movers   Movers that provide governance for adjusting generator power outputs. It is necessary to keep generator output in line with network load.

Protective relays Relays that shut down electrical systems or certain components under varying circumstances. Usually they consist of short circuits or abnormal currents that can interfere with or damage electrical equipment.

Proximity switches Switches that open or close circuits when the circuits get within a specific distance from another object.

Reactive power The resultant power loss derived from power generation. It is used for various purposes.

Rectifiers Converters that convert AC power into DC power and they may form part of a variable-speed system or a battery-charging system. Simple rectifiers use diode bridge circuits to convert AC to fluctuating DC.

Rotor (of a generator) A conductive metal cylinder that spins inside the stator. It is connected to the turbine rotor via the drive train. Its spinning action inside the stator allows for the conversion from mechanical energy delivered by the turbine rotor into electricity fed into the grid.

Rotor (of a turbine) The rotor of a turbine includes the blades and the hub. Normally, the rotor accounts for about 20 percent of the turbine's cost. Most turbines have either two or three blades that allow lift and rotation. The rotor turns the low-speed shaft at ~30 to 60 rotations per minute. The rotor speed is controlled by the blades that are turned (or pitched) out of the wind.

Rotor active power The amount of energy either supplied or removed by the voltage source converters. Rotor active power is derived from the following formula: $P_r = \frac{V_r}{s} I_r \cos\theta$.

Rotor dynamics A specialized field of study in wind energy. It is based on mechanical studies of torque and speed as they apply to the rotor and drive shaft.

Rotor flux magnitude and angle control scheme A doubly fed induction generator control scheme in which terminal voltage and power output are modulated. This happens via adjustment of the rotor flux vector's magnitude and angle.

Rotor flux vector A characteristic of electromagnetization. Generators that contain permanent magnets have fixed rotor flux vectors, but in electromagnetized generators, this vector can be changed. It is based on controlling stator currents and rotor slip.

Short-term wind forecasting An attempt to predict wind levels from 1 to 48 hours in advance. The best forecasting is based on models of current and past wind behavior for onshore and offshore developments.

Slip The ratio between synchronous speed and rotor operating speed.

Slow primary response Another form of occasional services. In this response, there is no need for immediate turbine response to load changes. The slight excess power available is driven to more transient loads. No turbine control system action is needed.

Solid-state relays (SSRs) Electric relays that contain no moving parts, thus extending the relay's life cycle. There are two major kinds of solid-state relays used in wind turbine operations: photo-coupled and transformer-coupled SSRs.

Squirrel cage generator A type of generator whose rotor does not contain coil windings. The rotor is made of two rings connected by conducting bars. The bars conduct electric currents and are covered by a laminate to protect them. The rotor's appearance is that of a "squirrel cage."

Spilling The name for the intentional reduction of wind power extraction. It leaves a margin for extra wind turbine loading.

Stator A round, static metal casing that contains copper windings or conducting rods. It functions as a canister that fixes the rotor in a parallel axis. This allows the rotor to rotate inside the round space via bearings or slip rings.

Time–delay relays   Relays that stay on for a certain amount of time once they are activated. They are made using adjustable timer circuits, which control the actual relay.

Tip-speed ratio (TSR)   The ratio between the blade tip speed and the current wind speed in a given moment. If the tip speed is exactly the same as the wind speed, tip-speed ratio (TSR) is 1. TSR design is one factor of wind turbine efficiency.

Total harmonic distortion (THD)   The sum of all harmonics emitted by a machine's distinct components.

Topology   The overall layout of the grid on the physical space it occupies. It includes wind patterns and electrical needs, which are partly based on regional geographical features.

Transmission substation   Substation that connects two or more transmission lines.

Vertical axis wind turbine (VAWT)   A wind turbine that operates in high-turbulence sites where wind direction is highly variable. The generator and gearbox are placed near the ground, so they do not require tall towers and high maintenance. There are three types of VAWTs: Darrieus, Giromill, and Savonius.

Voltage flicker   Sometimes referred to simply as *flicker*. This occurs when wind power generation fluctuates, causing visible light fluctuations in the connected grid.

Washout term   The upper limit on the gain that the differentiator output can have as the frequency of the input signal increases in a control system.

Wind   Air in motion, in any direction. As a type of solar energy, wind is caused by spatial differences in atmospheric pressure, specifically moving from high pressure to low pressure. Earth is unevenly heated by the sun; that is, the North Pole and the South Pole receive less solar energy than the equator, and dry land heats and cools quicker than the seas. This difference in heating causes global atmospheric convection—air flowing from high pressure to low pressure—which causes wind and the direction of wind is influenced by Earth's rotation.

Wind power density (WPD)   The calculation of effective wind force (velocity and mass) in a particular location, above ground level, over a specified time period.

Wind turbine   A wind-powered electrical machine that uses wind to drive an electric generator.

Wind vane   Outside the nacelle, the wind vane measures wind direction and communicates with the yaw drive to manipulate the turbine according to wind orientation.

Wound-rotor generator   A type of generator so named because rather than conducting bars, the generator has copper coil windings that transmit electrical currents.

Yaw   To turn to face the wind.

Yaw drive   The yaw drive keeps the rotor facing into the wind whenever wind direction changes; upwind turbines face into the wind, while downwind turbines require no yaw drive.

Yaw motor   The yaw motor powers the yaw drive.

References

20 Percent Wind Energy by 2030. (n.d.). *Increasing wind energy's contribution to U.S. electricity supply*. Retrieved August 8, 2010, from http://www.20percentwind.org/

Agrawal, A. (2006, September 15). Transformers pack powerful growth since '01. *Economic Times* (New Delhi). Retrieved December 28, 2010, from http://www.highbeam.com/doc/1G1-151446581.html?key=0142160D517E191168170E041A0E6F4B2E224E324D3417295C30420B61651B617F137019731B7B1D6B39

American Wind Energy Association. (n.d.). Retrieved August 8, 2010, from http://www.awea.org/pubs/factsheets.html

American Wind Energy Association. (2010). Annual Wind Industry Report. Retrieved February 1, 2012, from http://www.awea.org/learnabout/publications/reports/AWEA-US-Wind-Industry-Market-Reports.cfm

Anaya-Lara, O., Jenkins, N, Ekanayake, J., Cartwright, P., & Hughes, M. (2009). *Wind energy generation: Modelling and control*. West Sussex, England: John Wiley & Sons, Ltd.

Bellés, M. R. (2007). Drilling wind in the Middle East. *Vestas*. Retrieved August 9, 2010, from http://www.vestas.com/en/media/article-display.aspx?action=3&NewsID=1325

Blaabjerg, F., & Chen, Z. (2006). *Power electronics for modern wind turbines*. Copenhagen, Denmark: Aalborg University.

Bullock, M. (2010). How relays work: Relay construction. *How Stuff Works*. Retrieved November 3, 2010, from http://electronics.howstuffworks.com/relay1.htm

Burton, T., Sharpe, D., Jenkins, N., & Bossanyi, E. (2001). *Wind energy handbook, 1st ed.* West Sussex, England: John Wiley & Sons, Ltd.

California Institute of Technology. (2010, May 23). Schooling fish offer new ideas for wind farming. *Science Daily*. Retrieved December 18, 2010, from http://www.sciencedaily.com/releases/2010/05/100517152532.htm

Cooper, D. J. (2008). Integral action and PI control. *Practical process control e-textbook*. Retrieved April 28, 2011, from http://www.controlguru.com/wp/p69.html

Drye, W. (2010, April 28). First U.S. offshore wind project approved. *National Geographic*. Retrieved September 3, 2010, from http://news.nationalgeographic.com/news/2010/04/100428-energy-first-offshore-wind-project-approved/

United State Department of Energy. (2010). Wind and Water Power Program: *Supporting wind turbine manufacturing*.

Retrieved November 2, 2010, from http://www1 .eere.energy.gov/ windandhydro/

Eng-Tip Forums. (2004, December 9). *Variable frequency drives: How does vector control compare to servo control?* Retrieved December 31, 2010, from http://www.eng-tips.com/faqs .cfm?fid=1063

Environmental and Energy Study Institute. (October 2010). "Offshore Wind Energy." Retrieved January 11, 2012, from http://www.eesi.org/files /offshore_wind_101310.pdf

Green Energy Ohio. (n.d.). Retrieved August 10, 2010, from http://www .greenenergyohio.org/page.cfm? pageId=34

Hau, E. (2006). Wind turbines: fundamentals, technologies, application, economics. Berlin: Springer-Verlag. ISBN 3540242406

Magnet: Frequently-Asked Questions. (2011). *KJmagnetics.com*. Retrieved April 27, 2011, from http://www .kjmagnetics.com/FAQ.asp

Makower, J., Pernick, R., & Wilder, C. (2009). Clean Energy Trends 2009. Portland, OR: Clean Edge Inc.

Manwell, J. F., McGowan, J. G., & Rogers, A. L. (2009). *Wind energy explained: theory, design and application,* (2nd ed.). West Sussex, England: John Wiley & Sons, Ltd.

McKean, E. (ed.). (2010). *The new Oxford American dictionary* (2nd ed.). Oxford, England: Oxford Publishing.

Mechanics. (2010). Retrieved December 8, 2010, from http://www.sciencedaily .com/articles/m/mechanics.htm New Home Wind Power.com. *The history of wind power.* Retrieved August 10, 2010, from http://www.newhomewindpower .com/history-of-wind-power.html

Muhamad, N. D. (2010). Course lecture downloads: SEE 3433. *Universiti Teknologi of Malaysia.* Retrieved

April 28, 2011, from http://encon .fke.utm.my/courses/see3433/ lecture/2010-kl/dc_machine.pdf

NEMA Surge Protection Institute. (2009). *Glossary*. Retrieved May 25, 2011, from http://www.nemasurge.com/glossary .html

Nuclear Power Fundamentals: Motor Controller Types and Operation. (2010). *Integrated publishing.* Retrieved November 2, 2010, from http://www .tpub.com/content/doe/h1011v4/css /h1011v4_130.htm and http://www .tpub.com/content/doe/h1011v4/css /h1011v4_131.htm

Oracle ThinkQuest Education Foundation. (2010). *Periodic functions.* Retrieved December 28, 2010, from http://library .thinkquest.org/2647/algebra/ftperiod .htm

Plantier, K., & Smith, K. M. (2009). *Electromechanical principles of wind turbines for wind energy technicians.* Waco, TX: Texas State Technical College Publishing.

REN21 (2010). Renewables 2010 Global Status Report. Retrieved January 11, 2012, from http://www.ren21 .net/REN21Activities/Publications /GlobalStatusReport/GSR2010 /tabid/5824/Default.aspx

REUK.co.uk. (March 6, 2007). *Wind turbine tip speed ratio.* Retrieved September 4, 2010, from http://www.reuk.co.uk /Wind-Turbine-Tip-Speed-Ratio.htm

Roney, J. M. (2010). *Earth Policy Institute: Eco-economy indicators.* Retrieved May 25, 2011, from http://www.earth-policy .org/index.php?/indicators/C49/

Spera, D. (ed.). 2009. *Wind turbine technology: Fundamental concepts in wind turbine engineering* (2nd ed.). ASME Press, ISBN #: 9780791802601

Total Alternative Power. (n.d.). *Wind power. What is power?* Retrieved August 11, 2010, from http://www .totalalternativepower.com/faq.htm

Tsili, M., Patsiouras, Ch., & Papathanassiou, S. (n.d.). *Grid code requirements for large wind farms: A review of technical regulations and available wind turbine technologies.* Athens, Greece: National Technical University of Athens. Retrieved December 11, 2010, from http://users.ntua.gr/stpapath/Paper_2.72.pdf

Turbulence. (2010). *Science Daily.* Retrieved December 12, 2010, from http://www.sciencedaily.com/articles/t/turbulence.htm

United States Department of Energy (n.d.). *Wind and water power program.* Retrieved August 7, 2010, from http://www1.eere.energy.gov/windandhydro/index.html

United States Department of Energy. (2010, February 25). *Wind and water power program: Interconnecting renewable systems.* Retrieved January 7, 2011, from http://www1.eere.energy.gov/windandhydro/renewable_systems.html

United States Department of Energy. (2010). *Wind and water power program: Offshore wind technology.* Retrieved December 23, 2010, from http://www1.eere.energy.gov/windandhydro/offshore_wind.html

Wind energy 101. (2010). Power of wind: Powering a cleaner, stronger America. Retrieved November 20, 2010, from http://www.powerofwind.com/we_101.html

"Wind Energy Basics." (n.d.). American Wind Energy Association. From http://www.awea.org/

Wind Energy Development Programmatic EIS. (n.d.). *FAQs.* Retrieved August 10, 2010, from http://windeis.anl.gov/faq/index.cfm#Whatiswind

Woodlands Coalition. (2002). *Website glossary.* Retrieved May 25, 2011, from http://woodlandscoalition.com/Glossary.htm

Woodward Corporation. (1991). *Speed droop and power generation.* Retrieved May 1, 2011, from http://www.matsuda-gov.com/topic/droop&powergene.pdf

Index

Figures and tables are indicated by f and t following page numbers.